Contributions to Statistics

V. Fedorov/W.G. Müller/I. N. Vuchkov (Eds.)
Model-Oriented Data Analysis
XII/248 pages, 1992

J. Antoch (Ed.)
Computational Aspects of Model Choice
VII/285 pages, 1993

W. G. Müller/H. P. Wynn/A. A. Zhigljavsky (Eds.)
Model-Oriented Data Analysis
XIII/287 pages, 1993

P. Mandl/M. Hušková (Eds.)
Asymptotic Statistics
X/474 pages, 1994

P. Dirschedl/R. Ostermann (Eds.)
Computational Statistics
VII/553 pages, 1994

C. P. Kitsos/W. G. Müller (Eds.)
MODA4 – Advances in Model-Oriented Data Analysis
XIV/297 pages, 1995

H. Schmidli
Reduced Rank Regression
X/179 pages, 1995

W. Härdle/M. G. Schimek (Eds.)
Statistical Theory and Computational Aspects of Smoothing
VIII/265 pages, 1996

S. Klinke
Data Structures for Computational Statistics
VIII/284 pages, 1997

Christos P. Kitsos · Lutz Edler (Eds.)

Industrial Statistics

Aims and Computational Aspects

Proceedings of the Satellite Conference to the
51st Session of the International Statistical Institute (ISI)
Athens, Greece, August 16–17, 1997

With 44 Figures
and 24 Tables

Physica-Verlag
A Springer Verlag Company

Professor Christos P. Kitsos
Athens University of Economics and Business
Department of Statistics
76 Patission Street
104 34 Athens
Greece

Dr. Lutz Edler
German Cancer Research Center
Biostatistics Unit
Im Neuenheimer Feld 280
D-69120 Heidelberg
Germany

ISBN-13: 978-3-7908-1042-4 e-ISBN-13: 978-3-642-59268-3
DOI: 10.1007/978-3-642-59268-3

Industrial statistics: aims and computational aspects ; proceedings of the satellite conference to the 51st session of the International Statistical Institute (ISI), Athens, Greece, August 16 - 17, 1997 / Christos P. Kitsos ; Lutz Edler (ed.). - Heidelberg; New York : Physica-Verl., 1997
(Contributions to statistics)

Softcover Design: Erich Kirchner, Heidelberg

SPIN 10634241 88/2202-5 4 3 2 1 0 – Printed on acid-free paper

PREFACE

This volume is the Proceedings of the International Conference in Athens from August 16-17, 1997, on 'Industrial Statistics: Aims and Computational Aspects' planned and organized as a Satellite Conference to the 51th Session of the International Statistical Institute (ISI) in 1997 in Istanbul. Plans for this conference originated in Beijing in 1995, during the 50th Session of the ISI where active members of the ISI and especially the IASC (International Association of Statistical Computing) decided to organize such a meeting realizing the importance of statistics and its computational aspects in modern industry and economical development. Devoted to the growing impact of statistical methodology and statistical computing in industry and economy the aim of this conference was to link the three components Statistics-Industry-Computers towards a more intensive interaction which should improve industrial productivity, foster the impact of statistics in industry and promote there theory, methods and practice of statistical computing. According to this, the conference and the proceedings address themselves to decision makers, statisticians, computer scientists, programmers, engineers and other professionals in industry and academia engaged in this developing field.

The meeting was kindly cosponsored by the *International Association of Statistical Computing* (IASC) and the *Statistics in Industry Committee of the ISI.*

We take this opportunity to express our sincere thanks to Professores and Dres. Murrray Cameron (President of IASC), Jaromir Antoch and Antoine de Falguerolles (Chairmen of the Board of Directors of the European Regional Section of IASC), and Vijayan Nair (Chairman of the Statistics in Industry Committee of ISI) for their support in this effort. Further thanks go also to the Director of the ISI Dr. Zoltan Kenessey for his encouragement and to Daniel Berze and Allmut Hörmann for editorial assistance in the announcements of the conference in the ISI Newsletter and in the Statistical Software Newsletter (SSN), respectively.

The editors would also like to express their thanks to the Ministry of Education of Greece for their generous support. Our thanks are extended to the Turkish Organization Committee of the 51th Session of the ISI who kindly welcomed this meeting, especially Ambassadeur Orhan Guvenen. Warmest thanks go also to Athens University of Economics and Business and Dean Prof. A. Kintis for their welcome and help. Further thanks go to a number of people involved in the production of this volume. We have to thank for assistance in handling the texts which arrived in various electronical forms, especially Axel Benner and Lothar Pilz in Heidelberg and George Andraktas in Athens. We thankfully acknowledge the friendly and pleasant cooperation with Physica-Verlag. We thank especially Dr. Peter Schuster and Dr. Martina Bihn who set up the schedule for printing and monitored the publishing.

The target of the planning was to have all the papers reviewed and published before the meeting and therefore we had to impose tight time limitings and pressuring deadlines. The strict timely schedule might have hindered some of the applicants for participation at the conference to contribute and may have made others to decide to restrict themselves to an oral presentation only of their valuable work; we apologize for that. To those who worked hard for the review of all papers, the International Committee and the Local Committee, we would like to express sincere thanks. Most of the announcement work and the communication was succesfully operated through e-mail. This electronic process was also used for editing the papers and valuable time could be saved such that the target could be reached succesfully.

The dominant topic in the submission of papers was the Industrial Statistics which is a broad and dynamically growing field of statistical applications and usage of statistical computing and therefore some heterogeneity of the contributions submitted to the conference had to be expected. Nevertheless, at the end of the day, when we were happy to collect the papers a few categories had arizen very clearly which represent in our opinion cutting edge areas of industrial statistics each represented by a number of contributions. We finally categorized the submitted papers into the six categories

Quality Control, Engineering and Monitoring
Reliability and Failure Time Analysis
Experimental Design
Repeated Measurements - Multiple Inference
Pharmaceutical-Statistics
Computing, Imaging and Prediction

Admittedly, such a categorization was not an easy task at all and there remains subjectivity in this and even more in the classification of the single papers. Obviously, other important areas in industrial statistics are not so well represented in these proceedings by various reasons. Among those less represented ones are e.g. computationally intensive methods, Bayesian analyses, and more theoretical stochastic analyses, e. g. related to time series. Needless to say that there are fields of application e. g. in finance and insurance which could not been covered in this volume. Some of the topics missing here will be seen as oral contribution during the conference. On the other hand, we are content to have been able to bring together an illustrative spectrum of contributions to industrial statistics.

It is our hope that the present collection will prove to be of interest not only for the participants of this but also to a larger scientific community with interest in this particular and developing field of statistics.

Christos P. Kitsos Athens Lutz Edler Heidelberg

Contents

III. Experimental Design 149

VI. Computing, Imaging and Prediction 275

Part I

Quality Control, Engineering and Monitoring

Quality Engineering Using Robust Design and Analysis

SUNG H. PARK AND JAE J. KIM
Department of Statistics, Seoul National University
Kwanak-ku, Seoul, 151-742, Korea

Abstract

The concept of quality engineering is introduced, and the role of robust design is explained. Also the relationship between quality engineering and quality management is presented. Six steps of parameter design, which is the major part of robust design, are suggested, and an illustrated real case study at a Korean company is discussed to explain the six steps.

1 Introduction

In the manufacturing industry, activities concerned with quality can be divided into six stages: product planning, product design, process design, production, sales and after-sales service. It is important that company-wide activities are necessary to improve quality and productivity at each of these six stages. A company needs to build an overall quality system in which all activities interact to produce products of designed quality with minimum costs.

A product's cost can be divided into two main parts: before sale and after sale to the customer. The costs incurred before sale are the manufacturing costs, and the costs incurred after sale are those due to quality loss. A defective product which is scrapped or reworked prior to shipment is viewed as a manufacturing cost to the company, but not a quality loss.

Quality engineering (QE) is an interdisciplinary science which is concerned with not only producing satisfactory products for customers but also reducing the total loss (manufacturing cost plus quality loss). Hence, QE involves engineering design, process operations, after-sales services, economics and statistics. Taguchi's impact on the concept of

*Kitsos, C.P. and Edler, L., Eds.:Industrial Statistics, Physica-Verlag, Heidelberg, 1997

quality control in the manufacturing industry has been far-reaching. His QE system has been used successfully by many companies all over the world. He emphasizes the importance of designing quality control into the manufacturing processes. Also, he stresses that quality variation is the main enemy of QE and that every effort should be made to reduce the variation in quality characteristics. Taguchi extensively uses experimental designs primarily as a tool to design products more robust (which means less sensitive) to noise factors.

Robust design is an engineering methodology for optimizing the product and process conditions which are minimally sensitive to the various development and manufacturing costs. Taguchi's parameter and tolerance design are important tools for robust design. Taguchi asserts that product and process design have a much greater impact on product quality than manufacturing and inspection. Quality should be designed into the product and not inspected into it.

2 Quality engineering and quality management

2.1 Overview of quality engineering

Quality system can be divided into two parts as follows.

(1) Off-line quality system: Activities for quality of design through market research and product/process development, which are quality efforts away from production lines.

(2) On-line quality system: Activities for quality of conformance and quality of service through manufacturing care, inspection and customer service. These are quality efforts mainly on production lines.

During the product design and process design phases, there are three steps in each design phase. They are system design, parameter design and tolerance design. Their concepts are not explained here. For good explanations, see Taguchi(1986), Logothetis and Wynn(1989), Phadke(1989) and Park(1996). Quality engineering may be defined as "an effective system of engineering management to minimize the total loss of products to society through off-line and on-line quality activities". Figure 1 shows the major activities, methods and tools of quality engineering.

Figure 1. Overview of quality engineering

Quality engineering

Off-line quality system

Major activities	Product design Process design
Methods	System design Parameter design Tolerance design
Tools	Quality function deployment Design of experiments

On-line quality system

Major activities	Process design Product design
Methods	Process diagnostics Prediction and correction Measurement and action
Tools	Statistical process control tools Design of experiments

2.2 Stages of new product development

The stages of new product development are dependent on the type of product being developed. However, a general flow diagram for new product development is given in Figure 2. The stages may be divided into two parts: off-line and on-line quality systems. The off-line quality system starts with the survey of market information and ends with the set-up of managing methods for the manufacturing process. The on-line quality system begins with the production of sample products and ends with after-sales service activities. Note that three steps in design (system, parameter and tolerance) play a critical role in the off-line quality system. On the other hand, statistical process control (SPC) tools help keep manufacturing under control.

Figure 2. Stages of new product development and the quality system cycle

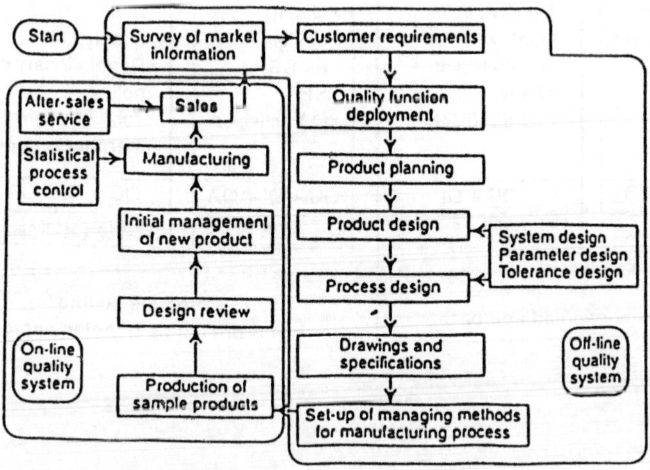

2.3 Quality management and quality engineering

Quality management (QM) is the overall management function that implements the quality policy (QP), quality control (QC), quality improvement (QI) and quality assurance (QA) for customer satisfaction under the leadership of senior management and with the full participation of all members of the organization. Since quality engineering (QE) is focused on product/process design and production for customer requirement, we may write QE and QM in the following equations:

$$QE = QC + QI + QA$$

$$\begin{aligned} QM &= QP + QC + QI + QA \\ &= QP + QE \end{aligned}$$

Figure 3 shows the development process of QM. Total quality control (TQC) is very similar to QM. However, the emphasis is different. In QM, senior management involvement and customer satisfaction carry more weight, but in TQC, QC and QI by line workers and middle managers are more important. Total quality management (TQM) is actually the same as QM. However, when we add T to QM, we mean to emphasize company-wide efforts for QM activities.

Figure 3. Development process of QM

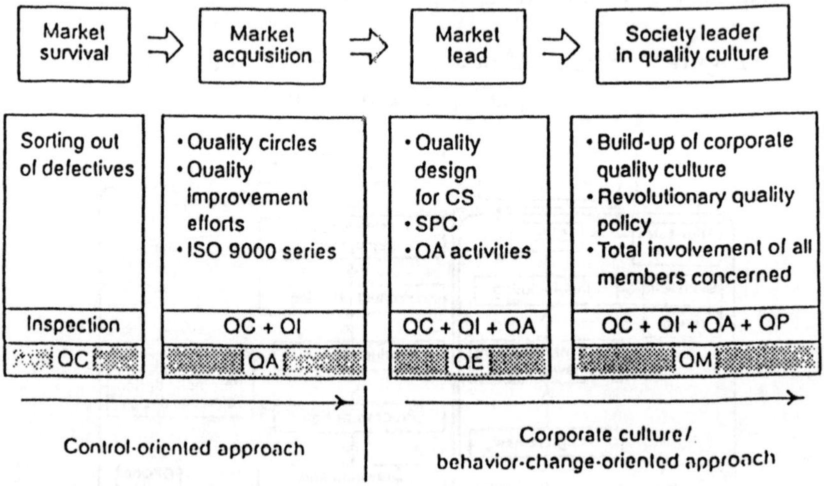

3 Parameter design

3.1 Steps of parameter design

Parameter design is the essence of robust design. It is usually difficult or very expensive to control or eliminate causes of variation which come from noise factors. Parameter design is used to improve quality without controlling or eliminating causes of variation, and to make the product robust against noise factors.

Six steps are needed for parameter design and data analysis, differing slightly according to the type of quality characteristic as follows.

(1) Smaller-the-better and larger-the-better characteristics:

(a) Sketch the factor-characteristic relation diagram for a given experimental problem.

(b) Select an appropriate orthogonal array (OA) for the inner array and another OA or a factorial layout for the outer array, and assign control and noise factors to the arrays.

(c) Conduct experiments, and compute the SN ratio for each row of the inner array.

Below n is the number of observation in each row.

(i) smaller-the-better characteristic

$$SN = -10log\left(\frac{1}{n}\sum_{j=1}^{n} y_j^2\right) \tag{1}$$

(ii) larger-the-better characteristic

$$SN = -10log\left(\frac{1}{n}\sum_{j=1}^{n} \frac{1}{y_j^2}\right) \tag{2}$$

(d) Find significant control factors which affect SN ratios through ANOVA (or pareto ANOVA). Find the optimum condition for the significant control factors.

(e) For the insignificant control factors, find the optimum condition based on other considerations such as economy, operability and easiness.

(f) Estimate the mean response at the overall optimum condition, and conduct a confirmation test at the optimum condition to check that the obtained optimum condition produces desired responses.

(2) Nominal-is-best characteristics:

(a)-(b) Same as above.
(c) Conduct experiments, and compute the SN ratio and sensitivity for each row of the inner array.

$$SN = 10log\left[\frac{1}{n}(S_m - V)/V\right] \tag{3}$$

or approximately

$$SN = 10log\left[\frac{1}{n}(S_m)/V\right] = 10log\left[\frac{\bar{y}^2}{V}\right] \tag{4}$$

$$\text{sensitivity} = 10log[S_m] = 10log[n(\bar{y}^2)] = 20log\left[\sqrt{n}\bar{y}\right] \tag{5}$$

where V is the sample variance, $\bar{y}$ is the sample mean, and S_m is the correction term,

$$S_m = \frac{1}{n}\left(\sum_{j=1}^{n} y_j\right)^2 = n(\bar{y})^2. \tag{6}$$

(d) Find significant control factors which affect SN ratios through (Pareto) ANOVA. Also find significant control factors which affect sensitivities through (Pareto) ANOVA. Classify the control factors into the following three categories.
(i) dispersion control factor: significant factors for SN ratios;
(ii) mean adjustment factor: significant factors for sensitivities;
(iii) insignificant factor: other control factors.
If a factor is both a dispersion control factor and a mean adjustment factor, it is classified as a dispersion control factor.
(e) Find the optimum condition as follows.

(i) dispersion control factor: the level which maximizes the SN ratio;

(ii) mean adjustment factor: the level which gives the closest estimated response to the target response value;

(iii) insignificant factor: the level which is chosen based on other considerations such as economy, operability and simplicity.

(f) Same as above.

3.2 An illustrated example

The six steps in parameter design which were discussed above are illustrated in this section with a real case study in a Korean electrical company. Here we will introduce a simplified ANOVA method which uses the Pareto principle. This method will be called **Pareto ANOVA**. This is a quick and easy method to analyze the results of parameter design which does not require an ANOVA table and, therefore, does not use F -tests.

An electrical company produces an electrical item called a ground fault circuit interrupter (GFCI), and the company was not happy with its high fraction defective for this product. It formed a task force to study the problem in 1995 and decided to carry out a parameter design to reduce the fraction defective. The quality characteristic was the moving time, which is a smaller-the-better characteristic. The factor-characteristic relation diagram shown in Figure 4 was produced.

Figure 4. Factor-characteristic relation diagram for GFCI experiment

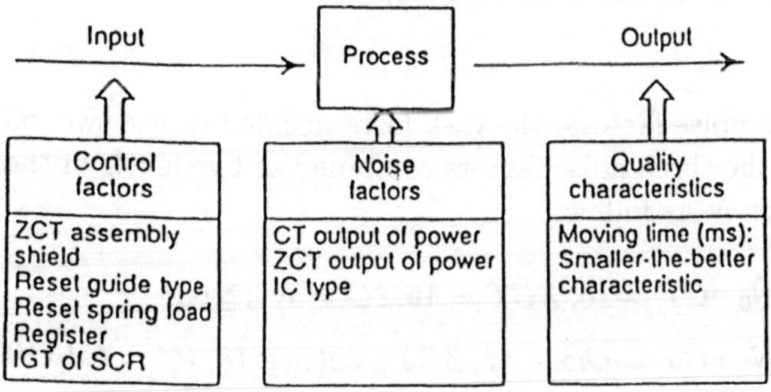

For the five control factors in Figure 1, the task force team decided to use two levels for each factor as follows, and assign the factors in L_8

as shown in Table 1.

Table 1 Experimental data and SN ratios for moving time

Array type	Inner array (L_8)							Outer array (one-way layout)						
Experiment number	Control factor assignment and column number							Raw data: moving time (ms)						SN= SN+23
	A	B	C	D	F	e	$C \times D$	Noise factor N						
	1	2	3	4	5	6	7	N_0			N_1			
1	0	0	0	0	0	0	0	7	14	16	11	17	10	0.73
2	0	0	0	1	1	1	1	8	14	13	12	16	12	0.90
3	0	1	1	0	0	1	1	13	11	13	8	12	16	1.13
4	0	1	1	1	1	0	0	11	13	13	10	14	13	1.12
5	1	0	1	0	1	0	1	17	14	14	13	15	12	−0.08
6	1	0	1	1	0	1	0	17	14	16	8	9	18	−0.05
7	1	1	0	0	1	1	0	13	10	13	12	13	13	1.14
8	1	1	0	1	0	0	1	8	11	9	13	10	12	2.46
Current condition								8	13	15	14	16	12	0.54

A : use of ZCT assembly shield (A_0= yes, A_1= no)
B : reset guide type (B_0= type I, B_1= type II)
C : reset spring load (C_0= 640g, C_1= 740g)
D : use of resistor (2.7 $k\Omega$) (D_0= yes, D_1= no)
F : electrical gate turn-on (IGT, μA) of (F_0= below, F_1= above
20)
silicon control rectifier(SCR)

For the noise factors, the task force decided to use two extreme conditions of the three noise factors combined as two levels of the combined noise factor N as follows.

$$N_0 : CT = 16, ZCT = 16, IC = KA2803$$

$$N_1 : CT = above\ 17, ZCT = above\ 18, IC = GL7101$$

At each level of the noise factor N, there are three observations for moving time. The current manufacturing condition for the control factors is

$A_1B_0C_0D_0F_1$ in which there are also six observations, as shown in Table 1.

The task force suspected that there might be some interaction between the reset spring load (C) and the resistor (D), so they assigned $C \times D$ to a column as in Table 1. For this parameter design, the six steps in the previous section will be roughly sketched.

The SN ratio for the ith row is computed using formula (1) since the quality characteristic is smaller-the-better. Note that there are six observations; hence, the SN ratio for the first row is

$$SN = -10log\left[\sum_{j=1}^{n} y_j^2/6\right]$$

$$= -10log\left[\left(7^2 + 14^2 + \cdots + 10^2\right)/6\right] = -22.27.$$

However, for computational simplicity, 23 is added to SN and it becomes 0.73 dB as written in Table 1. It should be recalled that if the dispersion is small and the average is small, the SN ratio becomes large. Thus the condition which makes the SN ratio largest is optimal. For many engineers and researchers in industry, ANOVA and related F-tests are not easy to use. Pareto ANOVA is a quick and easy methods for analyzing results. This method enables the significance of factors and interactions to be evaluated by Pareto-type analysis. It also allows the optimal levels of factors to be obtained. Table 2 shows the analysis for the GFCI experiment, from which we can see that the optimal levels of the control factors are $A_0B_1C_0D_1F_0$.

In Table 2 the square of the difference between the levels of A, for instance, is nothing but the numerator of S_A, since

$$S_A = \frac{(A_1 - A_0)^2}{8} = \frac{(3.47 - 3.88)^2}{8} = \frac{0.1681}{8}.$$

Also the contribution ratio of A, for instance, is computed as follows:

$$(0.1681/35.8519) \times 100 = 0.5\%$$

The factors and interactions of the shaded area in the Pareto diagram are considered to be significant sources of variation. Note that the significant factors and interactions are chosen from the left-hand side in the Pareto diagram, which cumulatively contribute about 90%. Since C × D is significant, we need to check the CD two-way table in order to find the optimal levels of C and D, and it turns out that C_0D_1 is optimal.

Comparing the sums at B_0 and B_1 factor levels, we can see that B_1 is optimal. Hence, the optimal combination of significant factor levels is $B_1C_0D_1$.

Table 2 Pareto ANOVA for two-level factors

Factor and interaction		*A*	*B*	*C*	*D*	*F*	*e*	*C* × *D*	Total
Sum at factor level	0	3.88	1.50	5.23	2.92	4.27	4.23	2.94	Sum of 0, 1 levels = 7.35
	1	3.47	5.85	2.12	4.43	3.08	3.12	4.41	
Difference at 2 levels		0.41	4.35	3.11	1.51	1.19	1.11	1.47	
Square of difference		0.1681	18.9225	9.6721	2.2801	1.4161	1.2321	2.1609	35.8519
Contribution ratio (%)		0.5	52.8	27.0	6.4	3.9	3.4	6.0	100.0
Pareto diagram		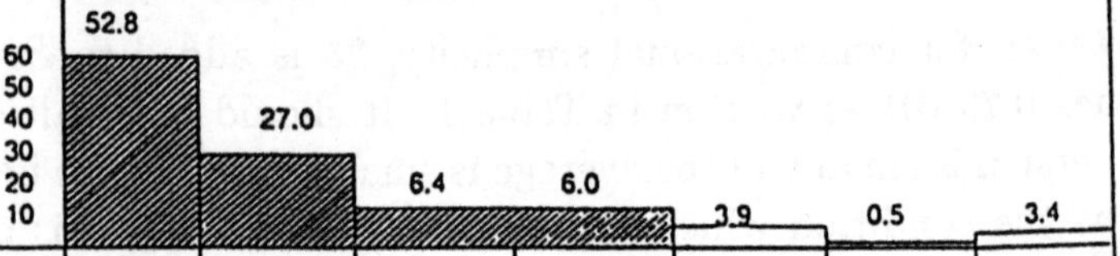							
Factor and interaction		*B*	*C*	*D*	*C* × *D*	*F*	*A*	*e*	
Cumulative contribution ratio (%)		52.8	79.8	86.2	92.2	96.1	96.6	100.0	

Check on significant interactions

CD 2-way table

	C_0	C_1
D_0	0.73 + 1.14 = 1.87	1.13 − 0.08 = 1.05
D_1	0.90 + 2.46 = 3.36	1.12 − 0.05 = 1.07

C_0D_1 is optimal.

Optimum combination of significant factor levels	$B_1C_0D_1$
Remarks on optimum conditions	The significant factors and interactions are chosen from the left-hand side in the above Pareto diagram which cumulatively contribute about 90%. If interactions are significant, the optimal levels of the related factors are chosen from two-way tables. For other non-significant factors, the optimal levels can be chosen from other economical and technical considerations.
Overall optimum conditions for all factors	A_1: ZCT assembly shield, yes B_1: Reset guide type, type II C_0: Reset spring load, 640 g D_1: Resistor, no F_0: IGT of SCR, below 20
Estimate of error variance	V_e = (pooled variation of non-significant sources)/(pooled degrees of freedom of non-significant sources) $= (S_A + S_F + S_e)/3$ $= (0.1681/8 + 1.4161/8 + 1.2321/8)/3 = 0.117$

If we choose A_0F_0 as non-significant factors for other economical and technical reasons, the overall optimal condition for all factors is $A_0B_1C_0D_1F_0$. At this condition the estimated SN ratio will be

$$\begin{aligned}\hat{\mu}(A_0B_1C_0D_1F_0) &= \bar{A}_0 + \bar{B}_1 + \bar{C_0D_1} + \bar{F}_0 - 3\bar{T} \\ &= \frac{3.88}{4} + \frac{5.84}{4} + \frac{3.36}{2} + \frac{4.27}{4} - 3 \times \frac{7.35}{8} \\ &= 2.42. \end{aligned} \tag{7}$$

For a smaller-the-better characteristic the mean squared deviation, the loss function and the SN ratio may be written as

$$\begin{aligned} MSD &= \frac{1}{n}\sum_{j=1}^{n} y_j^2 \\ L &= kMSD \\ SN &= -10log\,[MSD]\,. \end{aligned} \tag{8}$$

Let the MSDs of the current condition and the optimum condition be MSD_c and MSD_o respectively. Since the increase in SN from the current condition to the optimum condition is $2.42 - 0.54 = 1.88$,

$$\begin{aligned} 1.88 &= -10logMSD_o + 10logMSD_c \\ &= 10log\left(\frac{MSD_c}{MSD_o}\right). \end{aligned}$$

Hence,

$$\frac{MSD_c}{MSD_o} = 10^{0.188} = 1.54,$$

which means that, as compared with the current condition, the optimum condition produces the products with 35% less MSD. Also, from equation (8), the loss decreases by 35%.

Suppose the GFCI is judged defective when the moving time is above 20ms(i.e., $\Delta_0 = 20$), and the loss is $A_0^* = \$10$. Then the average loss per item for the current products becomes

$$\begin{aligned} L &= \frac{A_0^*}{\Delta_0^2} MSD_c \\ &= \frac{10}{(20)^2}\left[\frac{1}{6}(8^2 + 11^2 + 9^2 + 13^2 + 10^2 + 12^2)\right] \end{aligned}$$

$$= 0.025 \times 175.67 = \$4.39.$$

However, at the optimum condition the loss is decreased by a factor of 1.54, and the loss becomes

$$L = \frac{4.39}{1.54} = \$2.85,$$

which means that for each item the effect of quality improvement is $4.39 - 2.85 = \$1.54$. If the company produces 100000 items of GFCI, the total quality improvement effect amounts to \$ 154000 per year.

We obtained the estimated SN ratio in equation (7). If we subtract 23 (note that we added 23 to the SN ratios in Table 1) from this value, it becomes -20.58 dB. A 95% confidence interval for the average SN ratio at the optimum condition is

$$-20.58 \pm t(3, 0.025) = -20.58 \pm (3.182)\sqrt{\frac{V_e}{n_e}}$$

$$= -20.58 \pm 1.02 \qquad (9)$$

since the effective number of replication n_e is obtained from

$$\frac{1}{n_e} = \frac{1}{4} + \frac{1}{4} + \frac{1}{2} + \frac{1}{4} - \frac{3}{8} = \frac{7}{8}.$$

A confirmation test is conducted to check that the obtained optimum condition really produces the desired moving-time responses. The results are

$$12, 9, 8, 11, 10, 10, 9, 10, 11, 8.$$

The SN ratio for these observations is

$$SN = -10log\left[\frac{1}{10}(12^2 + 9^2 + \cdots + 8^2)\right]$$

$$= -19.89\text{dB}$$

Note that this value is contained within the 95% confidence interval obtained in (9), which means that the optimum condition is confirmed by a confirmation test.

ACKOWLEDGEMENT

This study was partially supported by Korean Ministry of Education through Research Fund, BSRI-96-1415.

REFERENCES

1. Logothetis, N. and Wynn, H.P. (1989) Quality through Design: Experimental Design, Off-line Quality Control and Taguchi's Contributions. Oxford, Clarendon Press.

2. Park, S. H. (1996) Robust Design and Analysis for Quality Engineering. London, Chapman & Hall.

3. Phadke, M.S. (1989) Quality Engineering Using Robust Design. Englewood Cliffs, N.J., Prentice hall.

4. Taguchi, G. (1986) Introduction to Quality Engineering. Tokyo, Asian Productivity Organization.

The Calibration Problem in Industry

C.P. KITSOS V.L. NINNI

Athens University of Economics and Business,
Patission 76, Athens 10434, Greece.

ABSTRACT

The calibration problem has been faced from different lines of thought (classical, Bayesian, structural, nonparametric, robust approach) and for different applications in industry (gas liquid chromatography, flame emission spectrometry, photometric analysis, etc.). The problem remains the evaluation of confidence intervals. This paper proposes an optimal design approach, adopting D and c-optimality, to obtain the asymptotic variance of the calibrating value, so that approximate confidence intervals can be evaluated.

Key-words: Calibration, regression, optimal design, photometric analysis, flame emission spectometry.

1 Introduction

Different methods and different models have been adopted in industry for the calibration problem. In practice most of the models attempt to face the simple linear calibration technique. Even when a non-linear model describes the phenomenon, Schwartz (1978) a linear approximation is eventually adopted. When the variance of the measurement is due to many sources of variability different techniques are followed, Schwartz (1979) while a robust approach might be appropriate, Kitsos and Muller (1995). Certainly to consider the variance constant and to follow a Statistical Quality Control method, Peroto and Converse (1991), might be helpful but not in all cases. Most of the problems in Industry adopt the regression method, Miller (1991) for the calibration problem and also. Frank and Friedman (1993).

*Kitsos, C.P. and Edler, L., Eds.:Industrial Statistics, Physica-Verlag, Heidelberg, 1997

The target of this paper is to discuss these methods and provide an alternative technique, based on optimum experiment design theory, Kitsos, (1992). So the experimenter can have a simple recipe when the simple calibration problem is to be applied.

2. Background

The calibration problem, at least in the simple linear case, has been discussed from different lines of thought: Classical, Bayesian, Structural, Non-parametric and as Robust calibration problem. For a compact review of the methods see Kitsos (1998) and for an extended discussion Osborne (1991).

Most of the applications in Chemistry, Schwartz (1978), Miller (1991), Sundberg (1988), deal with the classical approach, while Clark (1979, 1980) introduced a cross-validation approach.

In practice n readings are taken, y_i, i = 1, 2, ...n at predetermined known levels x_i, i = 1, 2, ...n. Then if a reading y_0, say, is observed it is asked to estimate the corresponding x_0 level. If we let $X = (x_0, x_1, \ldots x_n)$, $Y = (y_0, y_1, \ldots, y_n)$ the conditional distribution of X given Y cannot be evaluated from the conditional distribution of X given Y without data from the marginal distribution of Y corresponding to random X. For the Bayesian approach there is no such a problem as a prior for X' can always be assumed given. Moreover the relationship assumed between X and Y can be either linear or nonlinear.

In most of the practical applications a simple linear regression model is assumed, i.e.

$$y_i = \beta_0 + \beta_1 x_i + e, \qquad x_i \in T = [-1,1] \quad i=1,2,\ldots n \tag{2.1}$$

where $\beta = (\beta_o, \beta_1)' \in \Re \times (\Re - \{0\})$ is unknown parameters, $Z = (1,x)$ is the known regression function, and for the independent and identically distributed errors e is assumed that E(e) = 0. Var(e) = σ^2. Moreover the design space T is assumed to be without loss of generality the interval $[-1,1] \subseteq \Re$.

Given the favorite value y_0 the corresponding experimental condition x_0 should be estimated as:

$$x_0 = \frac{y_0 - \hat{\beta}_0}{\hat{\beta}_1} \tag{2.2}$$

The most well known competitive estimators of x_o when y_o is provided are the so-called "Classical Predictor"

$$C(x_0) = \bar{x} + \frac{S_{xx}}{S_{xy}}(\dot{y}_o - \bar{y}) \tag{2.3}$$

and the "Inverse-Predictor"

$$I(x_0) = \bar{x} + \frac{S_{xy}}{S_{yy}}(\dot{y}_o - \bar{y}) \tag{2.4}$$

with: $S_{tr} = \Sigma(t_i - \bar{t})(r_i - \bar{r})$

and by $\dot{y}_0$ we mean the possible average of the possible k observations taken at the prediction stage (or experimental condition) and $\bar{y}$ as usually.

The comparison of $C(x_0)$ and $I(x_0)$ is based on the values of n and $|\sigma/\beta_1|$ under the assumption that $x_0 \in T$ One of the merits of $C(x_0)$ is that when the usual normal assumption for the errors is imposed the classical predictor is the maximum likelihood estimator. Moreover $C(x_0)$ is a consistent estimator while $I(x_0)$ is inconsistent. The $I(x_0)$ estimation was critisized that provides confidence interval which might be the whole real line or in the best case two disjoint semi-infinite intervals, Graybill (1976). When $|\sigma/\beta_1|$ is small the asymptotic mse (mean square error) of $C(x_0)$ is smaller that the use of $I(x_0)$, when x_0 does not lie in the neighborhood of $\bar{x}$.

The main difficulty, the construction of confidence intervals, as the variance of x_0 does not exist, provides food for thought for an optimal design approach to the calibration problem, which we discuss in the next section.

3. Optimal design approach

Consider (2.2) as a non-linear aspect of β. In principle a non-linear aspect is a nonlinear function of the parameters of a linear model. That is $x_o = \varphi(\beta)$.

Technically speaking $Var[\varphi(\beta)]$ does not exist, however an asymptotic evaluation from (2.2) is

$$Var\,[\varphi(\beta)] \cong \beta_1^{-2} Z' Var(\hat{\beta}) Z \tag{3.1}$$

provided that $|\beta_1/\sigma|$ is not small. Now, if as an optimality criterion is adopted to choose that design which provides minimum variance i.e.

$$\min Var\,[\varphi(\hat{\beta})] \tag{3.2}$$

a D-optimal design is adopted. For a D-optimal design the following is true:

Proposition 1: For a D-optimal design it is

$$V_D(x_0) = (n\beta_1^2)^{-1}(x_0^2 + 1) \tag{3.3}$$

Proof:

If we let $\xi = \xi(p)$ to be the design measure allocating p observations at +1 and 1-p at -1 the corresponding design matrix $M = (m_{ij})$ i=1,2, j=1,2 has elements $m_{11}=m_{22}$-1 and $m_{12}=m_{21}=2p-1$ and therefore from (3.1):

$$Var\,\{\varphi(\hat{\beta})\} = (n\beta_1^2)^{-1} Z' M^{-1} Z$$

and eventually it is evaluated that

$$Var[\varphi(\hat{\beta})]=[4n\beta_1^2 p(1-p)]^{-1}\,[x_0^2+2x_0(1-2p)+1] \tag{3.4}$$

Then as the D-optimal design allocates half observation at the end points i.e. p=1-p=1/2 (3.4) is reduced to (3.3) Q.E.D.

Therefore a $(1-\alpha)100\%$ approximated confidence interval can be evaluated, when a D-optimal design is performed, as

$$CI(x_o) = (\,x_o - t_{n-1,\,\alpha/2}V_D(x_o),\;\; x_o + t_{n-1,\,\alpha/2}V_D(x_o)\,) \tag{3.5}$$

provided that β_1 has been replaced by an estimate. To obtain an estimate of β_1 the experimenter can devote a proportion of observation to get an estimate. This design is known as a two-stage design, Kitsos (1992).

This approximate confidence internal provides a solution, even when a Taylor approximation is considered, Schwartz (1978) due to influential observations. In such a case, i.e. when influential observations exist we would recommend a robust approach to the calibration problem, see Kitsos and Muller (1995).

4. Applications

Non-uniform variance of the calibration curve in quantitative chemical analysis is observed under some circumstances, Schwartz, (1979). Two of the methods used were the photometric analysis and the gas chromatographic analysis. In the former, although photometric analysis according to Beer's Law in normal conditions is expressed by a linear calibration line, in some cases there are deviations and the calibration curve has non-uniform variance. An example of this is the system aqueous copper (II) chloride, which in near ultraviolet has an absorption band deviating from Beer's Law, so the calibration curve is non-linear. In the latter, non-uniform variance can be observed in gas chromatographic analysis. In this case, a non-linear calibration curve appears when molar charges of benzene and solutions of benzene in toluence are injected. In most of the cases a linear approximation is suggested, so eventually a linear calibration problem is assumed.

But when the variance is not uniform the method adopted is not the appropriate. If with some probability, at a given experimental condition an outlying observation caused by a gross error may occur then the expected value of the error, E(e), is no longer zero. In such case, a conditionally contaminated linear regression model seems to be more appropriate to get estimators based on robust estimators for β. The robust linear calibration problem has been introduced by Kitsos and Muller (1995) and is applicable for such problems as Schwartz (1979).

Flame emission spectrometry (FES) is used to determine potassium, a very important element in Biology. There is a linear relationship between the concentration of potassium and the flame emission intensity. In spite of this, deviations are found in very large or small concentrations. A more reliable, simple statistical model can be used for the chemical analysis of FES by building a calibration curve. Spiegelman et. al. (1991). They are providing confidence interval for y_0 rather than x_0 under different assumptions for the variance. For the calibration problem it is difficult to calculate interval for x_0, rather to obtain "narrow" confidence intervals for y_0. Moreover they assume a bound for the variance and a robust approach might be more appropriate, Kitsos and Muller (1995).

Gas liquid chromatography is a method which is used in many circumstances in chemical analysis in research and industry. Chromatographic data is discussed by Kurtz (1983) who analyzed the collected data, and subjected it to gas liquid chromatography, through liquid chromatographic processes, with the MINITAB. He discusses the confidence interval problem for the fitted model, rather than for the x_0 value as we discussed in the above paragraph.

Quality control charts were used by Peroto and Converse (1991). They give emphasis that "collecting data and plotting charts is a futile act if nothing is done to improve the process", and we certainly agree. But to define UCL and LCL a more accurate procedure is needed. Therefore the two stage design might be useful: devoting a portion of the observations to get an estimate of β_1 and then evaluating the approximate variance (3.5) so that the appropriate control limits can be evaluated.

In the above four applications the confidence internal problem discussed in paragraph 3 appears to be crucial. The introduced confidence limits in (3.5) provide a useful approximation.

References

1. **Banks, D. (1993).** *Is Industrial Statistics Out of Control?* Statistical Science, 8, 356-409.
2. **Brown, P.J. (1982).** *Multivariate Calibration.* J.R. Statist. Soc. B 44, 287-321 (with discussion).
3. **Clark, R.M. (1979).** *Calibration, Cross-validation and Carbon-14.I.* J.R. Statist. Soc. A142, 47-62.
4. **Clark, R.M. (1980).** *Calibration, Cross-validation and Carbon 14.II.* J.R. Statist. Soc. A143, 177-194.
5. **Frank, I.E., Friedman, J.H. (1993).** *A Statistical View of Some Chemometrics Regression Tools.* Technometrics, 35, 109-148 (with discussion).
6. **Graybill, F. (1976).** Theory and Application of the Linear Model. Duxbury Press.
7. **Hochberg, Y., Marom, I. (1983).** *On Improved Calibrations of Unknowns in a System of Quality-Controlled Assays.* Biometrics, 39, 97-108.
8. **Hunter, W.G. (1981).** *Statistics and Chemistry, and the Linear Calibration Problem.* Chemometrics, Mathematics and Statistics in Chemistry, Kowalski, B.R. (ed.), 97-114.
9. **Kafadar, K. (1994).** *An Application of Nonlinear Regression in Research and Development: A Case Study from the Electronics Industry.* Technometrics, 36, 237-248.
10. **Kalotay, A.J. (1971).** *Structural Solution to the Linear Calibration Problem.* Technometrics, 13, 761-769.
11. **Kanatani, K. (1992).** *Statistical Analysis of Focal-Length Calibration Using Vanishing Points.* IEEE Transactions on Robotics and Automation, 8, 767-775.
12. **Kitsos, C.P., Muller, C.H. (1995).** *Robust Linear Calibration.* Statistics, 27, 93-106.
13. **Kitsos, C.P. (1992).** *Quasi-sequential procedures for the calibration problem.* In Compstat 1992, Y. Dodge and J. Whittaker (eds.), Physica-Verlag, 2, 227-231.

14. **Kitsos, C.P. (1998).** *Calibration.* In the Encyclopedia of Statistical Sciences, Update Vol. 3. S. Kotz, C. Read and D. Banks, eds. To appear.

15. **Kurtz, D. (1983).** *The Use of Regression and Statistical Methods to Establish Calibration Graphs in Chromatography.* Analytica Chimica Acta, 150, 105-114.

16. **Leary, J.J., Messick, E.B. (1985).** *Constrained Calibration Curves: A Novel Application of Lagrange Multipliers in Analytical Chemistry.* Analytical Chemistry, 57, 956-957.

17. **Lwin, T., Maritz, J.S. (1980).** *A Note on the Problem of Statistical Calibration.* Applied Statistics, 29, 135-141.

18. **Lwin, T., Maritz, J.S. (1982).** *An Analysis of the Linear-Calibration Controversy from the Perspective of Compound Estimation.* Technometrics, 24, 235-242.

19. **Lwin, T., Spiegelman, C.H. (1986).** *Calibration with Working Standards.* Applied Statistics, 35, 256-261.

20. **Merkle, W. (1983).** *Statistical Methods in Regression and Calibration Analysis of Chromosome Aberration Data.* Radiation and Environmental Biophysics, 21, 217-233.

21. **Miller, J.N. (1991).** *Basic Statistical Methods for Analytical Chemistry. Part 2. Calibration and Regression Methods. A Review.* Analyst, 116, 3-14.

22. **Naes, T., Irgens, C., Martens, H. (1986).** *Comparison of Linear Statistical Methods for Calibration of NIR Instruments.* Applied Statistics, 35, 195-206.

23. **Osborne, C. (1991).** *Statistical Calibration: A review.* International Statistical Review, 59, 309-336.

24. **Oman, S.D., Wax, Y. (1984).** *Estimating Fetal Age by Ultrasound Measurements: An Example of Multivariate Calibration.* Biometrics, 40, 947-960.

25. **Pevoto, L.F., Converse, J.G. (1991).** *Decisions to Change Analyzer Calibration Based on Statistical Quality Control Charts.* ISA Transactions, 30, 79-91.

26. **Schwartz, L.M. (1978).** *Statistical Uncertainties of Analyses by Calibration of Counting Measurements.* Analytical Chemistry, 50, 980-985.

27. **Schwartz, L.M. (1979).** *Calibration Curves with Nonuniform Variance.* Analytical Chemistry, 51, 723-727.

28. **Sjostrom, M., Wold, S., Lindberg, W., Persson, J., Martens, H. (1983).** *A Multivariate Calibration Problem in Analytical Chemistry Solved by Partial Least-Squares Models in Latent Variables.* Analytica Chimica Acta, 150, 61-70.

29. **Smith, J. (1990).** *Statistical Aspects of Measurement and Calibration.* Computers ind. Eng.. 18. 365-371.

30. **Spiegelman, C., Watters, R., Hungwu, L. (1991).** *A Statistical Method for Calibrating Flame Emission Spectrometry which Takes Account of Errors in the Calibration Standards.* Chemometrics and Intelligent Laboratory Systems, 11, 121-130.

31. **Srivastava, V., Singh, N. (1989).** *Small-Disturbance Asymptotic Theory for Linear Calibration Estimators.* Technometrics, 31, 373-378.

32. **Sundberg, R. (1988).** *Interplay between Chemistry and Statistics, with Special Reference to Calibration and the Generalized Standard Addition Method.* Chemometrics and Intelligent Laboratory Systems, 4, 299-305.

33. **Vecchia, D., Iyer, H., Chapman, P. (1989).** *Calibration with Randomly Changing Standard Curves.* Technometrics, 31, 83-90.

34. **Walters, F., Rizzuto, G.T. (1988).** *The Calibration Problem in Statistics and its Application to Chemistry.* Analytical Letters, 21(11), 2069-2076.

CASE STUDY OF APPLYING STATISTICAL TECHNIQUES FOR THE QUALITY ASSURANCE IN A PAINT INDUSTRY

C. Vourvahi, J. Voutsinas, J. Kiriakidis

CHROTEX S.A., 19300 Aspopyrgos, Greece

1. Introduction

Nowadays, under the term of *Quality Control* the following procedures are included:

- Evaluation of the actual quality performance.
- Comparison of actual performance to quality goals.
- Taking action on the difference.

Through such a system, the aim is always the Quality Assurance. There are many tools that enable the people of industry to attain this goal. Three of the most important ones are the Statistical Process Control (**SPC**), the Statistical Quality Control (**SQC**) and the Design of Experiments (**DOE**). The definitions of the above ones are given briefly.

SPC: The application of statistical techniques for the measurement of process variability

SQC: The application of statistical techniques for the measurement and improvement of the process quality. The SQC combines SPC, diagnostic tools, sampling plans, and other statistical techniques.

DOE: Generally, it is the construction of a complete plan for the conduction of experiments and evaluation of their results [1]. In the case of our field, the implied experiments are those that are needed for the recognition of the causes of deviations and their resolution.

The sequence between the above tools is demonstrated in figure 1.

*Kitsos, C.P. and Edler, L., Eds.:Industrial Statistics, Physica-Verlag, Heidelberg, 1997

In this paper, there will be no reference to DOE phase, but the SPC system that has been applied by our company - a paint industry - will be described, together with the results of this system to the quality of our products. The presentation will be supplemented by specific examples from our experience.

2. Generally about SPC

The general principles of SPC define that for its right implementation, the following steps must be made [2]:

- Recognition of the control requirements.
- Definition of variables, attributes[1] and parameters.
- Definition of specifications for the above.
- Establishment of documents and procedures for the collection and treatment of data.
- Nomination of SPC observers and co-ordination of the whole organisation to analyse and respond in time at the ''out of limits'' trends of the process.
- Preparation of periodic reports SPC.
- Suggestion of weaknesses and potential actions for improvement.

3. SPC in CHROTEX until 1996

Generally

In the next paragraphs, the operation practice of SPC in CHROTEX will be presented and examples from our experience until today, will be exhibited. This SPC system was implemented for the first time in 1994, and in 1997 it was enriched with new techniques (control charts, capability process indexes).
Until 1996, all the statistical techniques that were included in the SPC system, were mostly used for:

I. The control of production process and precision of the applied formulas.
II. The controls and adjustments which are carried out during the production process.
III. The evaluation of the results from the quality control of the final products.

All the above help the Technical Manager of the firm to check the compliance with the predetermined specifications and to establish new control limits for the product quality level.

Through SPC system, the existence of certain problems became clear and well stated. This many times lead to the formulation of solutions and corrective actions.

Thus, through the statistical observation of the product quality, the systematic

[1] *Variable* is a countable property, whereas *attribute* is a binary property, e.g. the presence of sediment in the paint.

deviations of some properties from their specifications became obvious in specific products. After recognising these deviations, various factors that may had influenced the product were re-examined. Such factors could be the formulation of the product, the used raw materials (especially if the prime materials themselves had respective deviations from their specifications), the method for measuring the property, and other special factors accordingly to the products nature (phase of DOE).

Our aim in this field is to advance to more complex statistical techniques, which will be incorporated in the present SPC system for the first time during this year, desiring always the highest Quality Assurance.

Collection of data

A basic presupposition for the statistical analysis, is the determination of measurable properties. This is the reason for which the technical quality specifications of our products have as much as possible countable magnitudes and not binary properties. With the data from measurements, which are executed accordingly to standard methods and preferably to ELOT, ISO, DIN, ASTM, etc. When a sufficient number of data for a product family or for a single product has been gathered, then one or more of the following statistical tools are used.

Statistical tools

1. Quality Control Table

All the measured technical characteristics are sorted by product and the average, the minimum value, the maximum value and the standard deviation for every characteristic are calculated.

2. Histogram

The measurements distribution is recorded for each of the properties and product. This graph is indicative of the fluctuation degree that a property of a product has.

3. Time Dependent Chart

The measured values of each property are presented versus the examined batches in time.

Annual Quality Report of Production

At the end of each year, the Annual Quality Report of Production is edited by the Technical Manager, and it includes:

1. An integrate table of final products quality. This table includes.
   a. Product code
   b. Counter of the produced batches during the year.
   c. Sum of kilos from standard formulations.

d. Sum of kilos from raw materials.
e. The degree of deviation from standard formulation.
f. Sum of produced kilos.
g. The production loss.
h. Upper limit of technical specifications.
i. Lower limit of technical specifications.
j. Mean value of specifications.
k. Minimum measured value.
l. Maximum measured value.
m. Average measured value.
n. Standard deviation of the measured values.

Comparison of all the above with respective values of previous year.

2. Integrate table of quality results for each of the group of 100 basic products of the company.
3. Histogram of technical characteristics for each of the group of 100 basic products of the company.
4. Time dependent chart for each of the group of 100 basic products of the company.
5. Results from the annual inspections of the group of 100 basic products of the company.
6. Conclusions that come out after the study and analysis of all the above elements. These conclusions concern the evaluation of the quality level that has been attained by the firm, the need for imposing corrective actions, and the definition of quality improvement targets for the forthcoming year.

4. New steps for CHROTEX with SPC

The SPC system that was described above, has already applied in CHROTEX giving important information for the quality level of products, for the systematic deviations from the specifications, for the steadiness or unsteadiness of a process. In 1997, new features and statistical tools were added in our techniques:

- ✓ control charts and
- ✓ capability process indexes.

Both of these tools are known to the most industrial people who work with statistics. For this reason I will briefly present them to you and continue with other topics.

Control charts: Establishing trial control limits

Control charts for data on variables are used to show the measurements in time sequence. The $\overline{X}$ chart shows the variability in group averages which may indicate a lack of control relative to an overall average value due to an assignable cause, and the Range chart which, when in control, measures the random or common cause variation in the process.

The data are arranged in time sequence and summarised in *rational subgroups*. The *underlying concept* of control charts is that within a subgroup there is only chance, random, or common cause variation; while all assignable or systematic variation will occur between subgroups [4].

A surprising number of companies try to use $\overline{X}$ and R charts for product acceptance purposes. They form subgroups by taking individuals at random from the entire population of measurements. This is an incorrect procedure. Subgroups must be made in production sequence [5]. In CHROTEX, we have defined subgroups of 2 measurements, that represent the two different points that every batch is measured.

The following procedures are being done in preparing the $\overline{X}$ and R charts [8]:

1. For each subgroup of 2 measurements (called a sample), the average, $\overline{X}$, and the range, R, are calculated.
2. Calculation of the grand average of all the samples, $\overline{\overline{X}}$, and of the average range, $\overline{R}$ of all the samples.
3. Calculation of the upper and lower control limits (UCL and LCL) using the equations:

$$UCL = \overline{\overline{X}} + \frac{3}{K \cdot \sqrt{n}} \cdot R \tag{1}$$

$$LCL = \overline{\overline{X}} - \frac{3}{K \cdot \sqrt{n}} \cdot R \tag{2}$$

   where K is a constant depending on the subgroup size (for size of 2, K=1.128), and n is the subgroup size (for our case n=2).

4. Combining these limits and by adding the line of grand average, $\overline{\overline{X}}$, in a chart which is constructed by plotting the subgroup averages horizontally in time sequence. The result of this is the $\overline{X}$ chart. An example is given to the following figure 2.

5. Calculation of the upper and lower control limits for the R chart using the equations:

$$UCL = C_1 \cdot \bar{R} \tag{3}$$

$$LCL = C_2 \cdot \bar{R} \tag{4}$$

6. Again the constants C_1 and C_2 are dependent on the subgroup size. In our case it is $C_1=0$ and $C_2=3.27$.
7. Combining these samples with the samples and by adding the line of the range average, $\bar{R}$, the R chart is produced. An example is given in the figure 3.

Drawing conclusions from the 2 charts.

When we say the ''process is out of control'' what we really mean is that there is variation present which is not just experimental error, i.e. there is something present in the system which is varying the data beyond realistic expectation of error. Also, when we say the ''process is in control'' (all the points fall inside the control limits) what we really mean is that for practical purposes, there is no significant cause of variation except experimental error. We cannot say that ''there are no causes of variation except experimental error'' as no statistical test can give a positive assurance; they can only give negative assurances.

Even though all of the points lie between the control limits, lack of control may also be shown by having a number of points lying on the same side of the average line in succession. Other type of runs of points may also indicate a lack of control. Further statistical steps may be performed to investigate this.

The control limits are designed to be, if the process is under control, ±3ό, from the average, so that the spread of the control limits is 6ό. Any lack of control shown in the $\overline{X}$ chart, is an indication of some assignable cause and must be investigated. But it is not sufficient to examine only the $\overline{X}$ chart, even if it shows control. Control limits for the $\overline{X}$ chart are meaningful, if the R chart shows lack of control.

If the R chart displays lack of control, we try to see from all pertinent information what may have caused this lack of control. If these points which indicate a lack of control are as a result of unusual conditions, they can me removed from the R chart. If they are just as a result of usual practices (points can lie outside the control limits, as they are in effect 99.7% confidence levels) they should not be removed. After the points that can be removed are, we calculate new control limits and examine if there is lack of control. If there is, then we examine the production process and impose corrective actions [3, 4, 8].

5. Process capability indexes

Process capability is a measurable property of the process (much as volume in litres is a measurable property of a container). The resulting measure is expressed in terms of 6ó and is unrelated to the product tolerance, i.e., ''the process does not know what the tolerance is''. However, the capability measurement is compared to the tolerance in order to judge the adequacy of the process.

Thus, a process capability index is utilized to answer the question: ''Is the available process capable of satisfying client 's requirements?' Such a process capability index is the C_p, where:

$$C_p = \frac{\text{Client 's requirements}}{\text{Process variability}} \tag{5}$$

which means:

$$C_p = \frac{USL - LSL}{6\sigma} \tag{6}$$

where, USL: Upper Specification Limit, and LSL: Lower Specification Limit.

We should always have in mind, that calculation of process capability index is meaningful, only when the process is ''in control'', which means that it has stable and predictable behaviour, i.e. it is described by a normal distribution.

In reality, index C_{pk}, as previously defined, leads frequently to wrong conclusions about the capability of process, because it does not take into account the position of the average (centring). There is a better capability index, named C_{pk}, which is defined like this:

$$C_{pk} = \min\left\{\frac{\text{distance of average, } \overline{X}\text{, from specification limits}}{1/2 \text{ process spread}}\right\} \tag{7}$$

or:

$$C_{pk} = \min\left\{\frac{USL - \overline{X}}{3\sigma}, \frac{\overline{X} - LSL}{3\sigma}\right\} \tag{8}$$

When we find that $C_p > 1.0$, it seems from the equation (6) that the spread of measurements is less than the area between specification limits. However, we cannot be sure of the capability because we have no indication of average, $\overline{X}$, versus limits. Index C_{pk} can give more information for the process capability, because it shows the transposition of the average versus a specification limit. The client or the final product user is interested only for index C_{pk}, at least at a first phase.

On the other hand, the first priority for a process engineer is index C_p, given that if C_p is not greater than a predetermined lower limit (usually 1.33), he/she does not go on with designing the process. In the case that C_p is close or lesser than 1.0, then his/her first action is to improve C_p and then to perform all the necessary moves of the process, in order to achieve an acceptable C_{pk} (figure 4) [1, 2, 3].

6. Applications of SPC

In the next examples, the implementation of SPC until today and its results in particular products, will be demonstrated. At this point, it should be pointed out that each year 's production results are treated during next year. So, in 1997 (this year), the production of 1996 has been evaluated.

Product A

In 1995 (figure 5a), great differences between different measurement points existed. Also, there was a period that quite many batches were out of specification at the final stage (QC) although they were OK during first control measurement. The histogram (figure 5b) shows a big spread which goes beyond specification limits.

After making these remarks, Technical Manager together with the relative chemist decided certain corrective actions and the examination of some factors which mainly were: prime materials, product stability between production and packing phases.

In the next year, we used firstly the same charts and compared with the ones from 1995 production. What we saw is a small improvement. This is obvious from the histogram (figure 5d) which indicates that a bigger part of the process than in 1995 is within specifications. However, there are still values outside specification limits, and it is also seen from the time dependent chart (figure 5c).

The next step we made, was the introduuction of control charts and capability indexes, as it was previously stated. In the $\overline{\overline{X}}$ chart (figure 6a) there was no sample at all outside control limits and in the range chart (figure 6b) there was only one. Thus, we concluded that the process was mainly in control.

Then we calculated the capability process indexes. C_p came out to be 0.612 and C_{pk} was 0.325 (i.e. $0<C_{pk}<1$). This means that a part of the process is not capable or that we should change the specifications.

After the evaluation of all the above, the conclusion is this:

The production of product A in 1996 was generally improved from 1995, and the process seemed to be in control. However, considerations about changing specification limits or part of the process were made.

Product B

This case is about a product having a property which is difficult to stabilise. During 1995, we had no exact specification limits, because they were being changed from period to period. However, the product seemed to have a narrow spread (figure 7b).

In 1996, specifications limits were established. These limits were established without using control charts, but by using only the theoretically expected values. It is very clear (figures 7c, 7d), that the process did not respond to these theoretical specification limits. The deviations were caused by systematic and not random errors. The histogram shows a normal distribution of the measured property, but a not acceptable spread of the process.

Accordingly to the theory of control charts, we constructed $\overline{X}$ and R charts (figure 8). We also calculated C_p and C_{pk} indexes. The conclusion was: *The process was almost in statistical control (UCL-LCL $\cong 6\sigma$), but all of the process is not capable of agreement with specifications given that the average $\overline{X}$ is out of limits.*

This case is still under study and methods of DOE are implemented aiming either to establish specification limits relative to process or to moderate the process so that it fulfills the specifications.

Product C

In 1996 the spread of the process was smaller than in 1995 (figure 9), although there were still values outside the specification limits. After constructing the control charts (figure 10), we noticed that the process was in control and that the samples overcoming the control limit in range chart were due to random errors.

The calculation of capability index gave for C_p the value 1.22 and for C_{pk} the value 0.26. The conclusion for this case was that: *The process is capable but it should be closely observed because a part of it does not satisfy the specification.*

We are now considering to make the next step towards control charts of crucial characteristics of the process that effect the quality of the product. The use of flow charts and cause-and-effect diagrams for the process can help to identify the key process variables for control and aid in the proper selection of subgroups.

7. Conclusions

The SPC system implemented until today has demonstrated its effectiveness and proved to be a useful tool in locating systematic deviations as well as separating them from random errors. Through SPC the recognition of problems and the imposing of corrective actions became easier. The awareness of existed deviations lead to re-examination of various parameters which could have affected the product.

Such factors could be the recipe, the utilized prime materials, the measuring method of the property, and other more specific parameters depending on the nature of the product.

In a progressive atmosphere, new products, process and product modifications, as well as manufacturing difficulties can lead to periods of increased non-conforming material. It is at these times that statistical process control procedures are vital to obtain minimum process capability and as a vehicle for process improvement. And it is precisely at these times that acceptance control procedures can act as a bulwark to prevent defective material from passing through and to allow the process to continue to operate to achieve the full benefit of process control [6].

Before ending, I would like to add that SPC demands increasing levels of sophistication in progressing toward high levels of capability and control. It is through this knowledge that the suppliers will ultimately build quality into the product while containing cost.

REFERENCES

1. Juran 's Quality Control Handbook, 4th edition, J.M.Juran, 1962, McGraw-Hill editions.
2. Δ. Γεωργι'ου, "SPC: τι ει'ναι και σε τι χρησιμευ'ει", διη'μερο - Διασφα'λιση Ποιο'τηταζ: ISO 9000 και η εφαρμογη' του στην Ελλα'δα, Τεχνικα' Χρονικα', 1/96, σσ. 92-104.
3. Β. Χριστο'ττουλοζ, "Στατιστικο'ζ ε'λεγχοζ ποιο'τηταζ σε χημικε'ζ βιομηχανι'εζ συνεχου'ζ παραγωγη'ζ", διη'μερο - Διασφα'λιση Ποιο'τηαζ: ISO 9000 και η εφαρμογη' του στην Ελλα'δα, Τεχνικα' Χρονικα', 1/96, σσ. 79-84.
4. Locke, John W., ''Statistical Measurement Control'', Quality and Statistics: Total Quality Management, ASTM STP 1209, Milton J. Kowaleski, Jr. Ed., American Society for Testing and Materials, Philadelphia, 1994.
5. Barry H. Levine, Duncan C. McCune, ''All you ever wanted to know about control charts: But never asked'', Quality and Statistics: Total Quality Management, ASTM STP 1209, Milton J. Kowaleski, Jr. Ed., American Society for Testing and Materials, Philadelphia, 1994.
6. E. G. Schilling, ''The transition from sampling to SPC'', Quality and Statistics: Total Quality Management, ASTM STP 1209, Milton J. Kowaleski, Jr. Ed., American Society for Testing and Materials, Philadelphia, 1994.
7. A. Barba, C. Anton, ''Applying Statistical Process Control'', Eurocoatings Journal, 3/1993.
8. Paint Research Association, ''Statistical Quality Control for CHROTEX S.A.'', Ref. 96/134/MP, 1996.

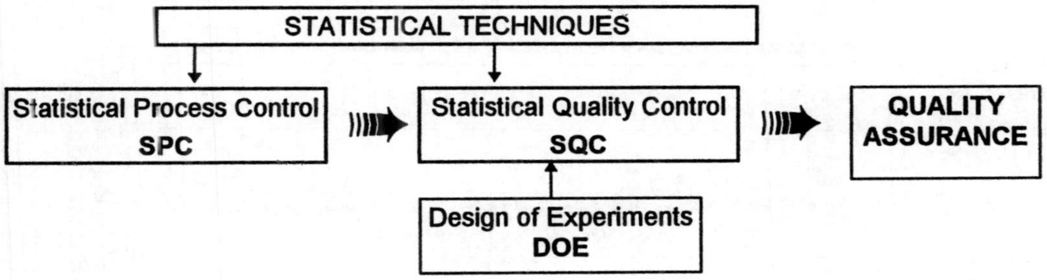

Figure 1: Aiming quality assurance.

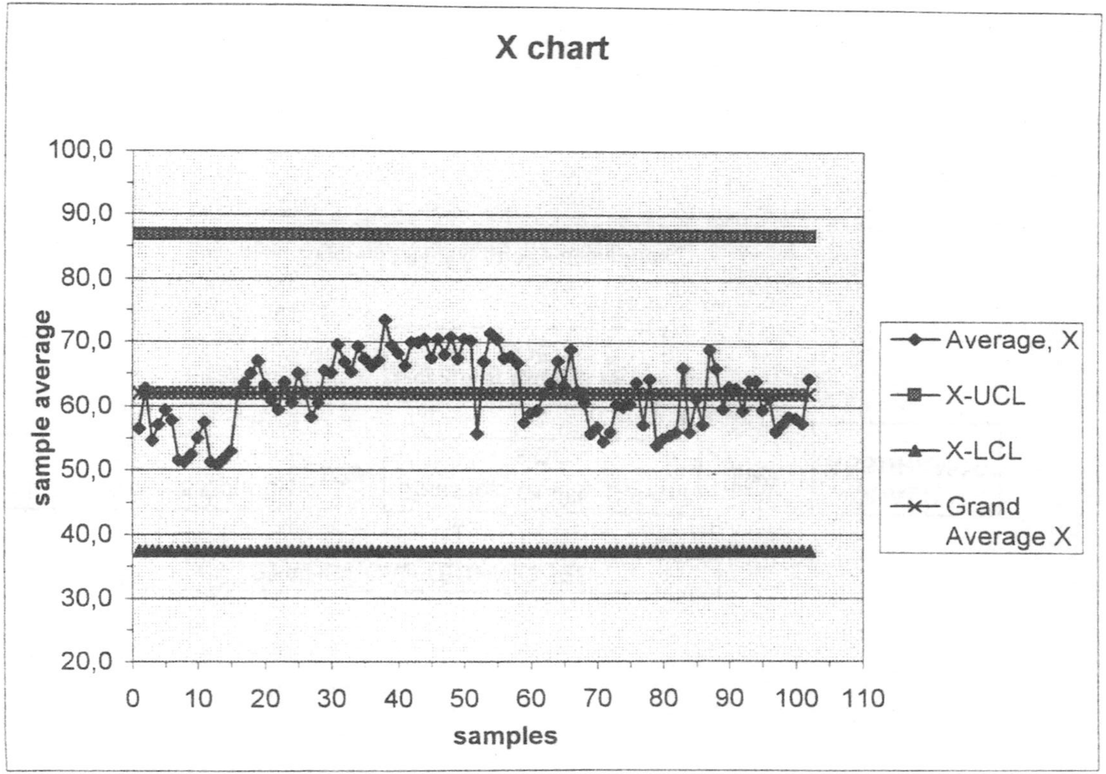

Figure 2: Example of X Chart.

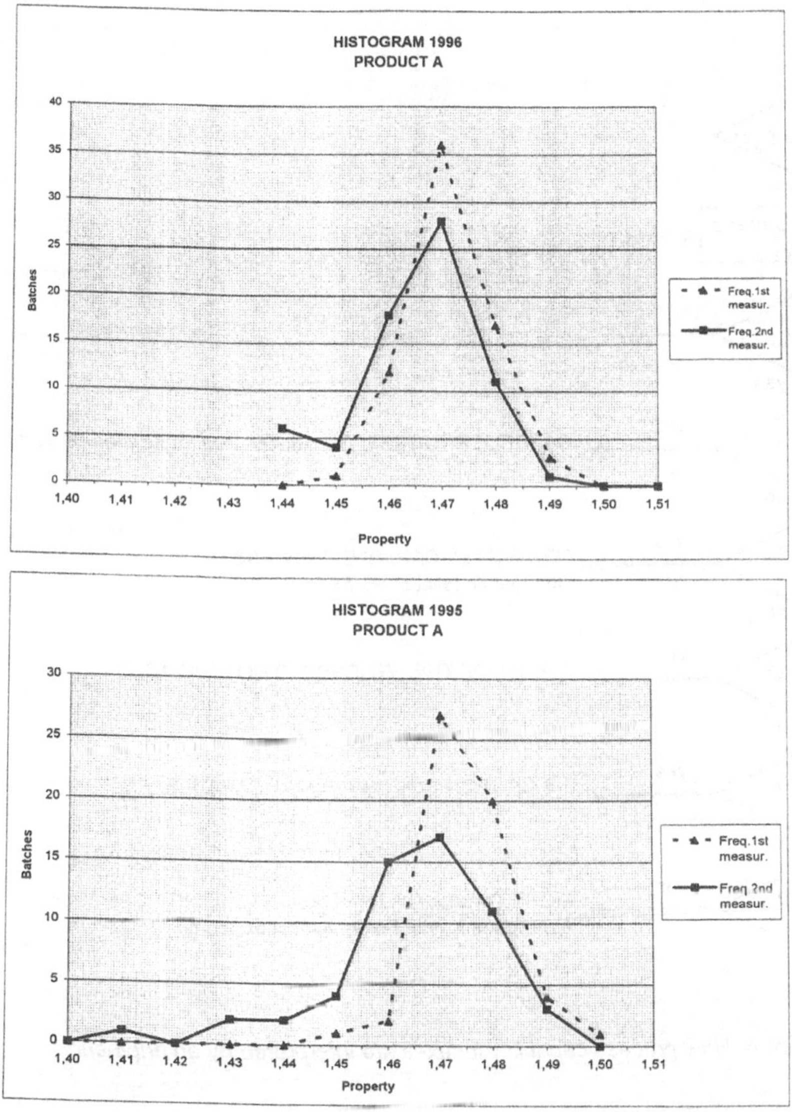

Figure 3: Example of R chart.

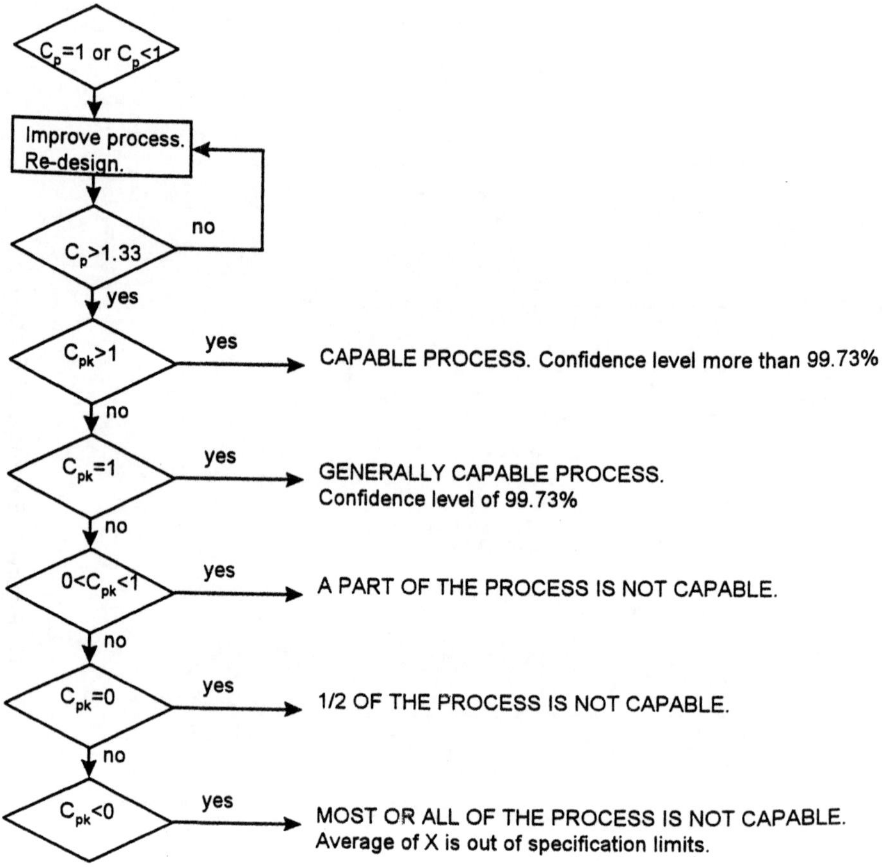

Figure 4: How process capability indexes are interpreted by an engineer.

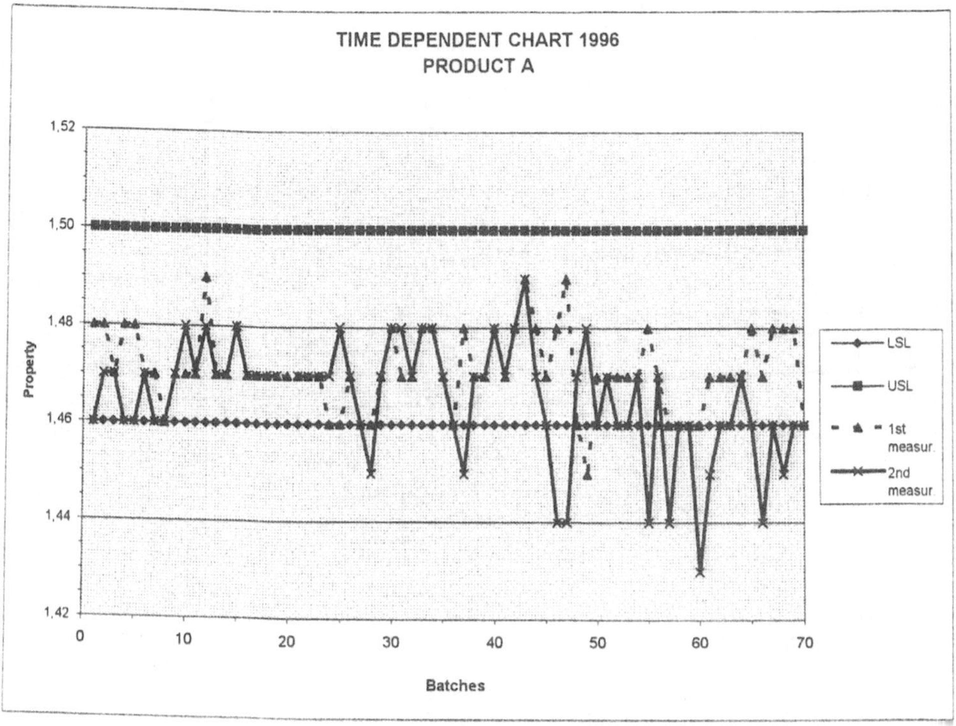

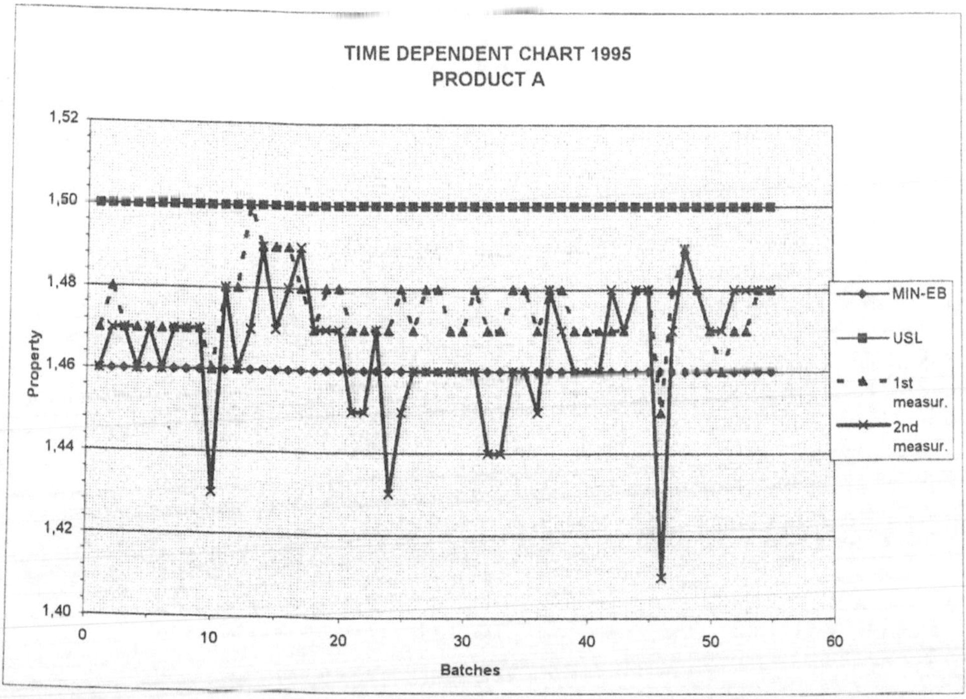

Figure 5: Product a - Comparison of 1996 production to 1995

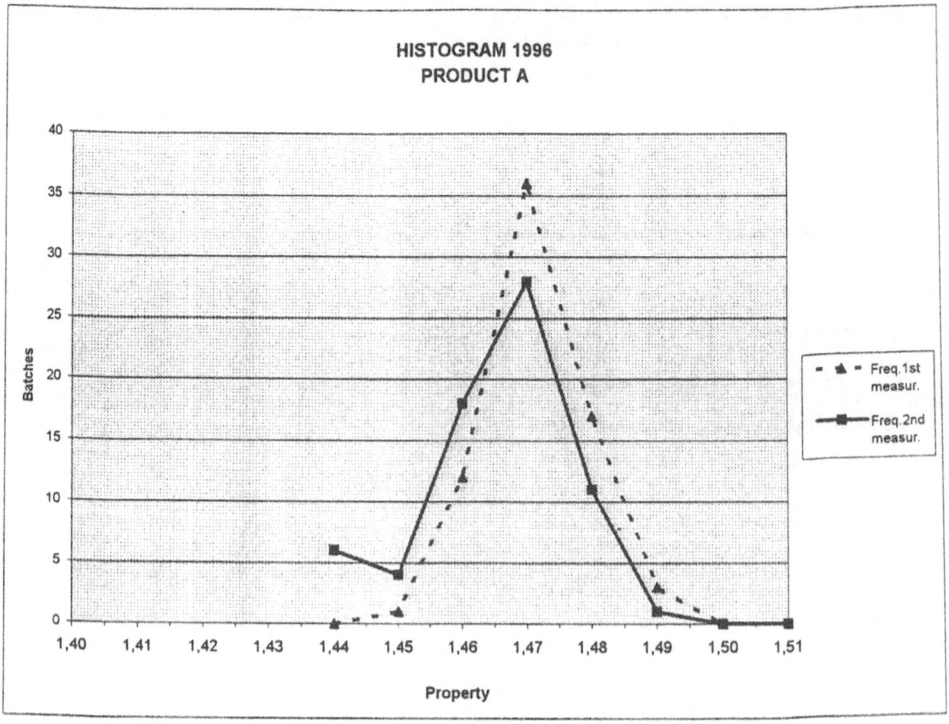

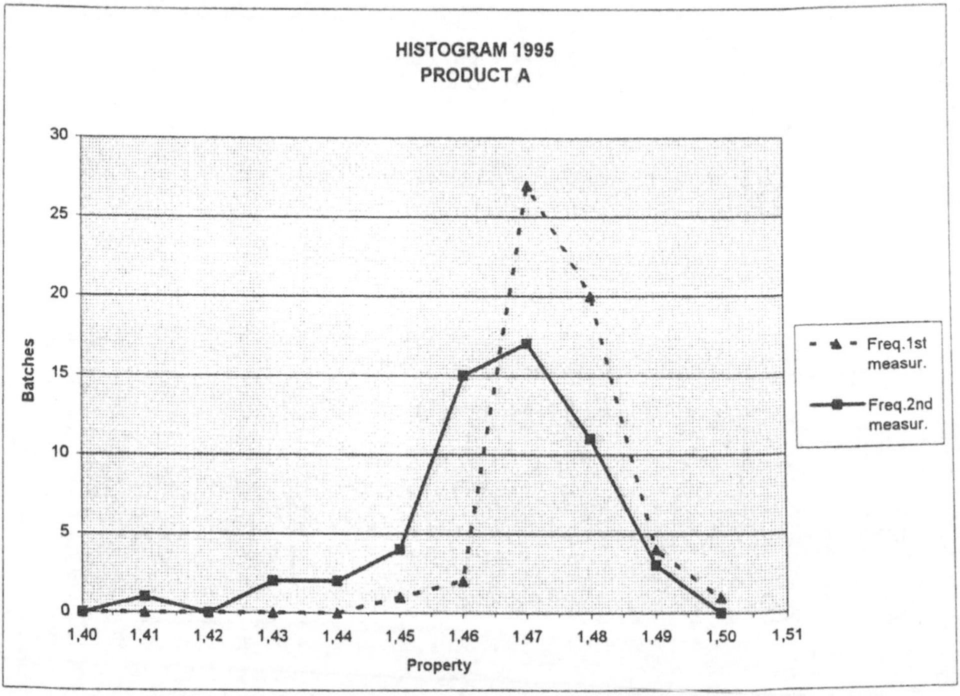

Figure 5: Product a - Comparison of 1996 production to 1995

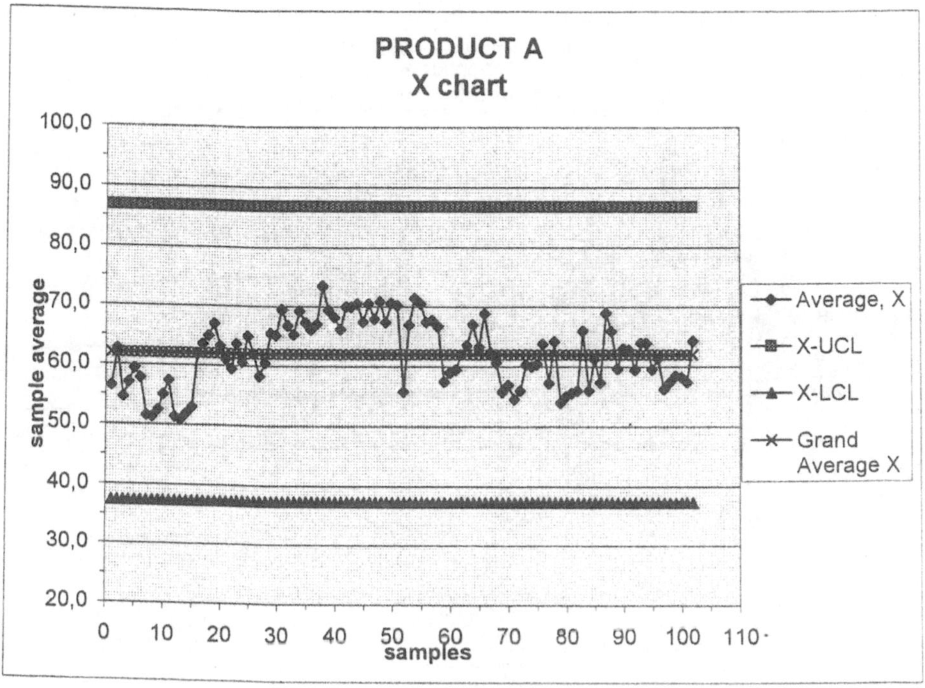

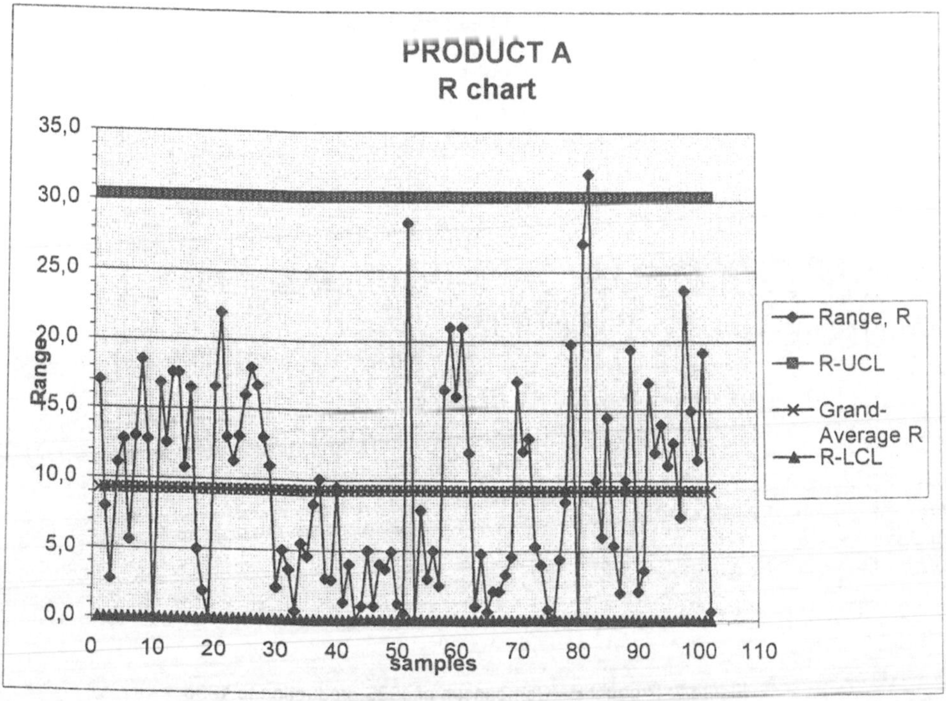

Figure 6: Product A - Control charts

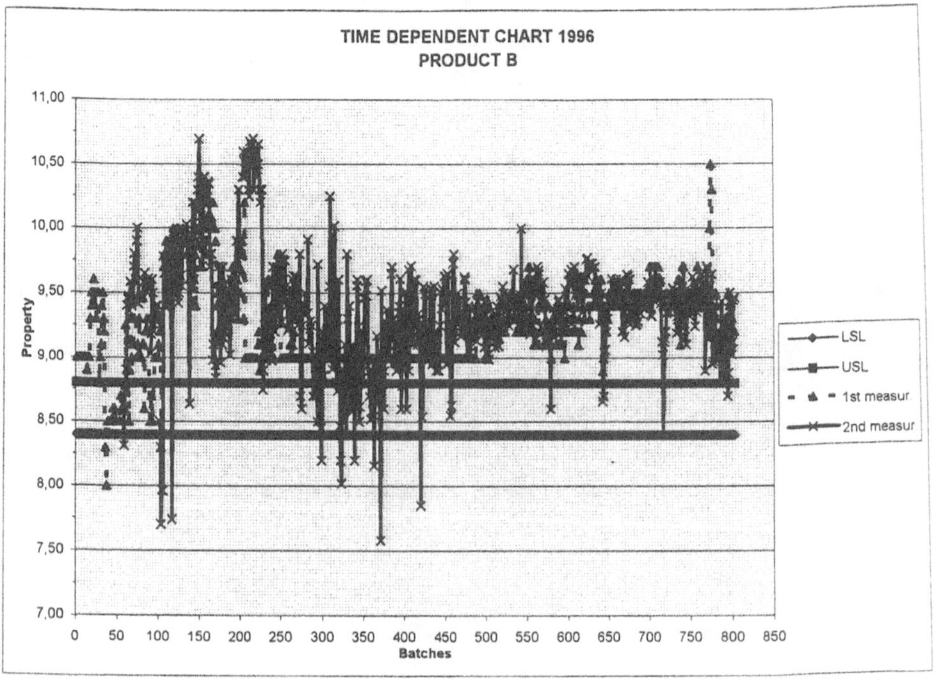

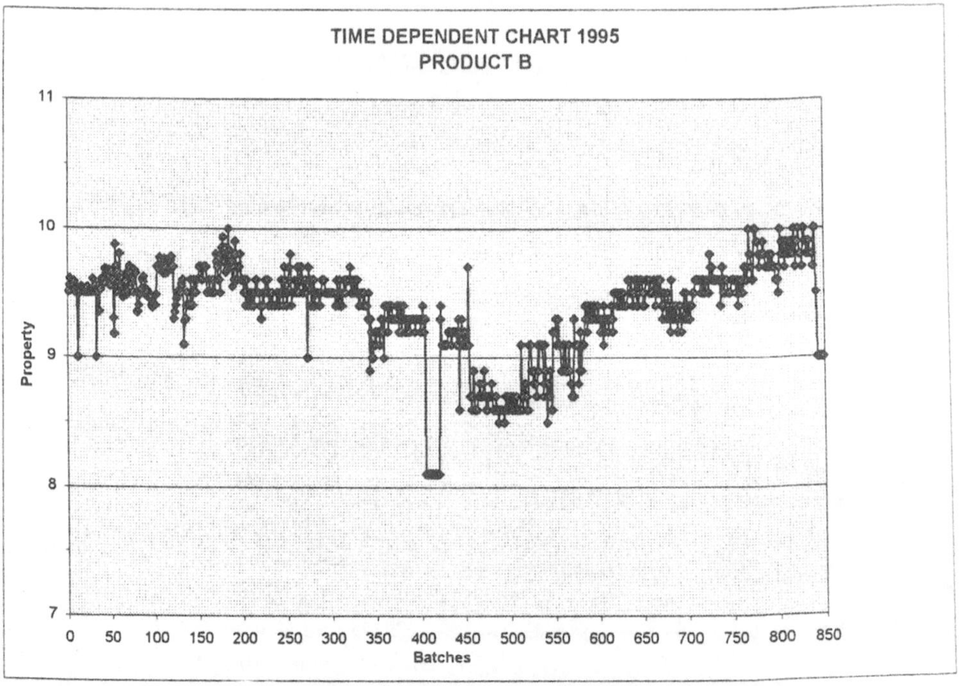

Figure 7: Product B - Comparison of 1996 production to 1995

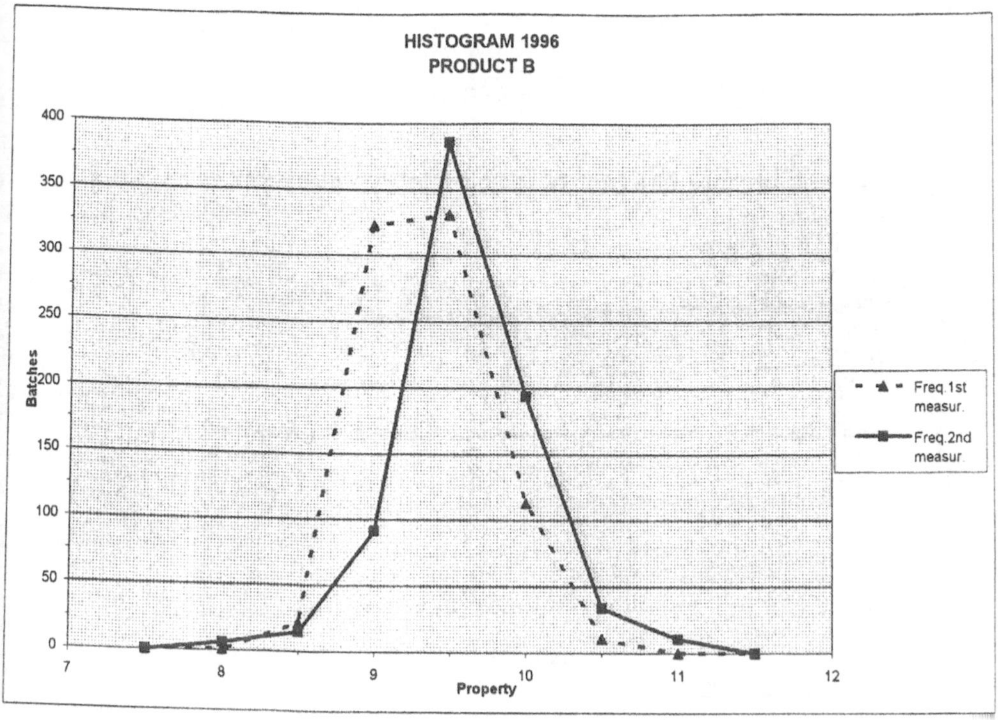

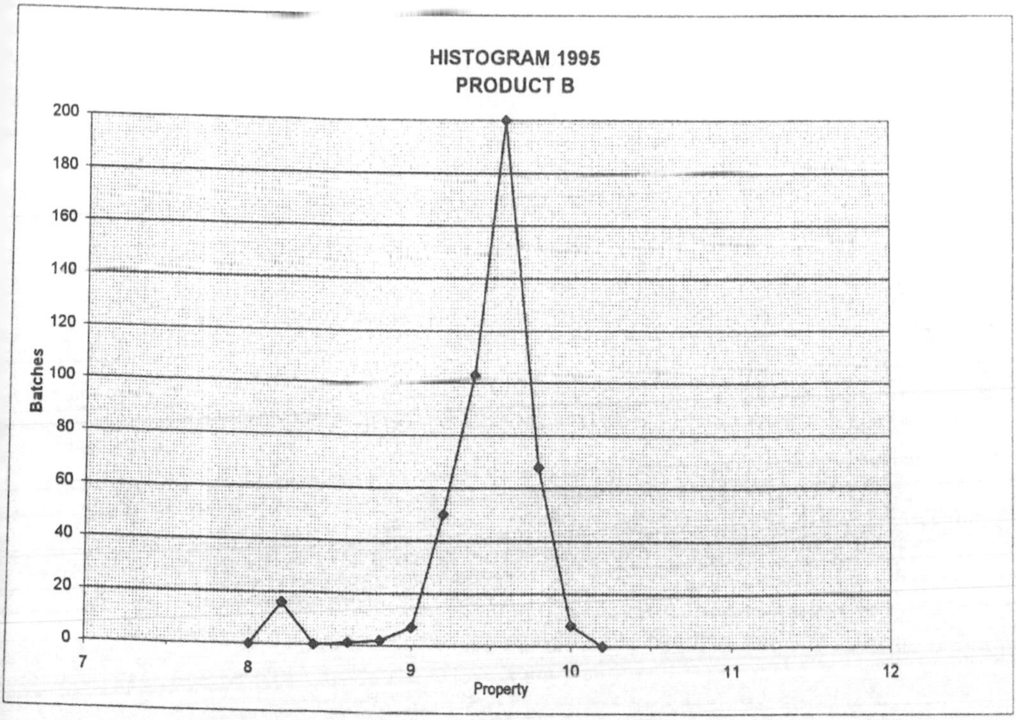

Figure 7: Product B - Comparison of 1996 production to 1995

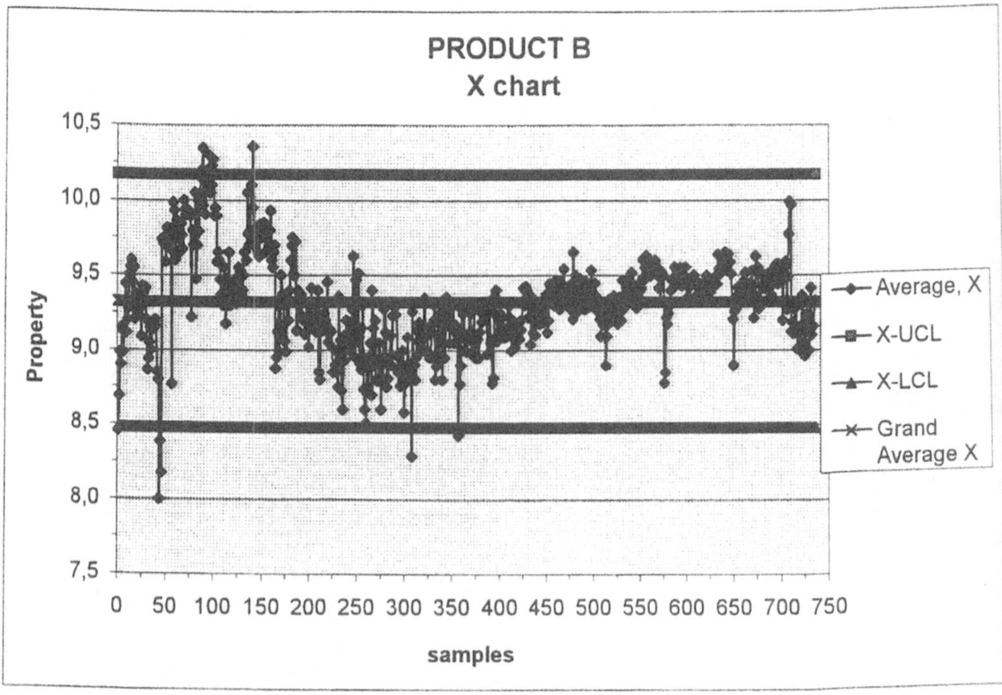

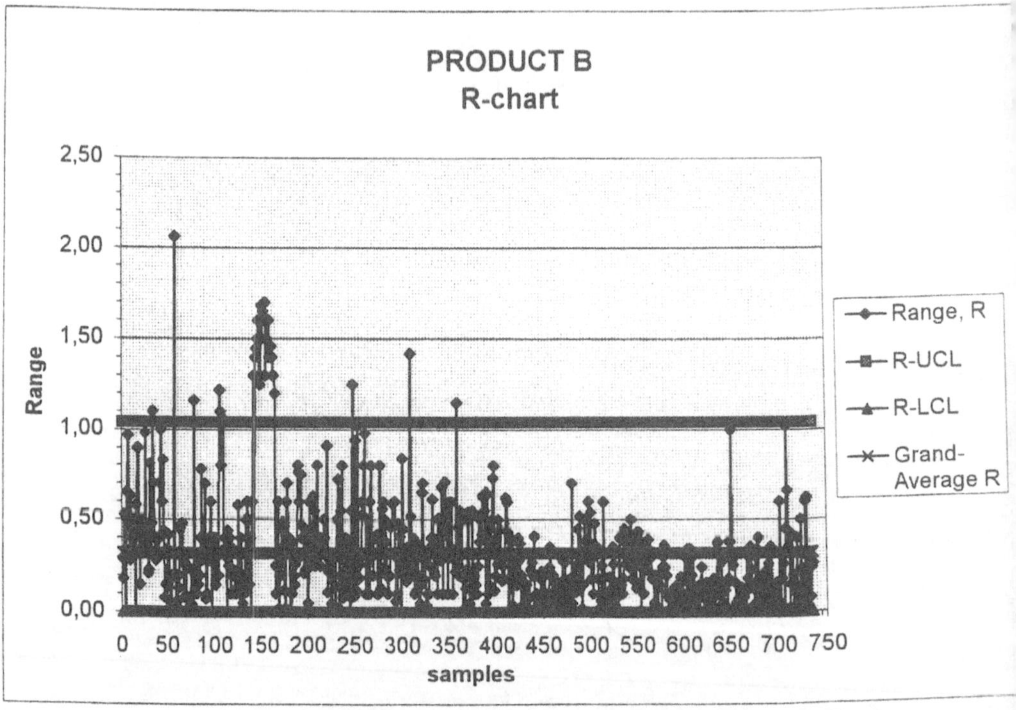

Figure 8: Product B - Control charts

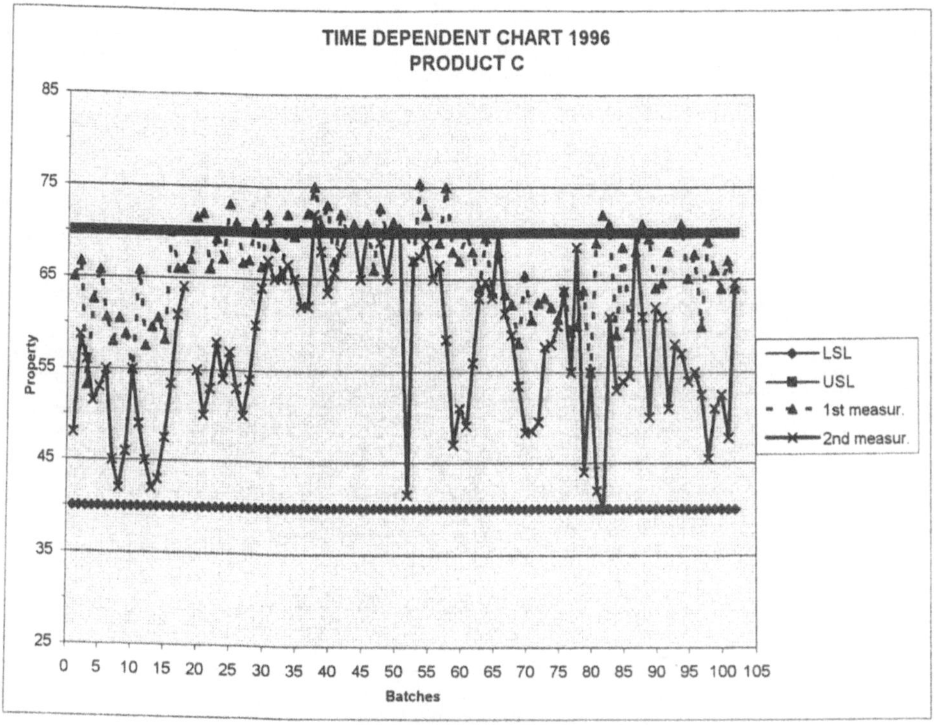

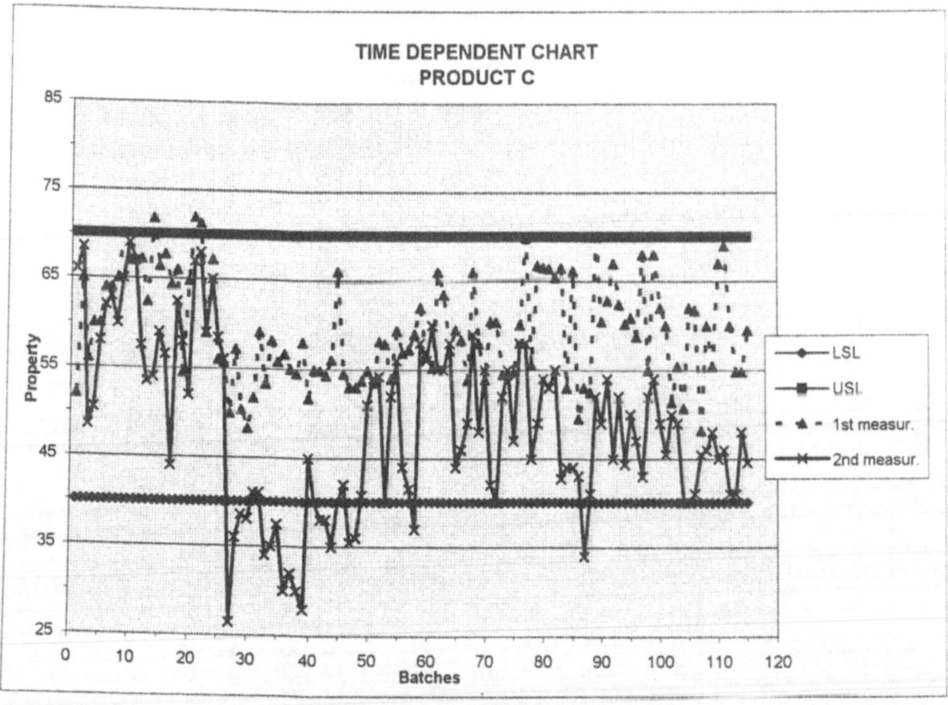

Figure 9: Product C - Comparison of 1996 to 1995.

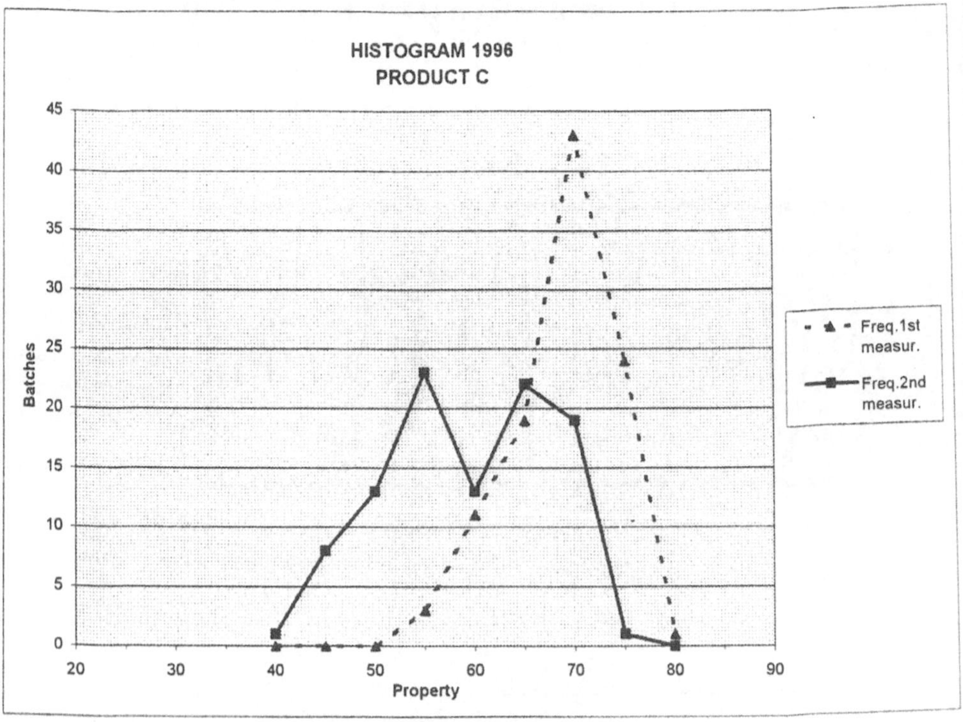

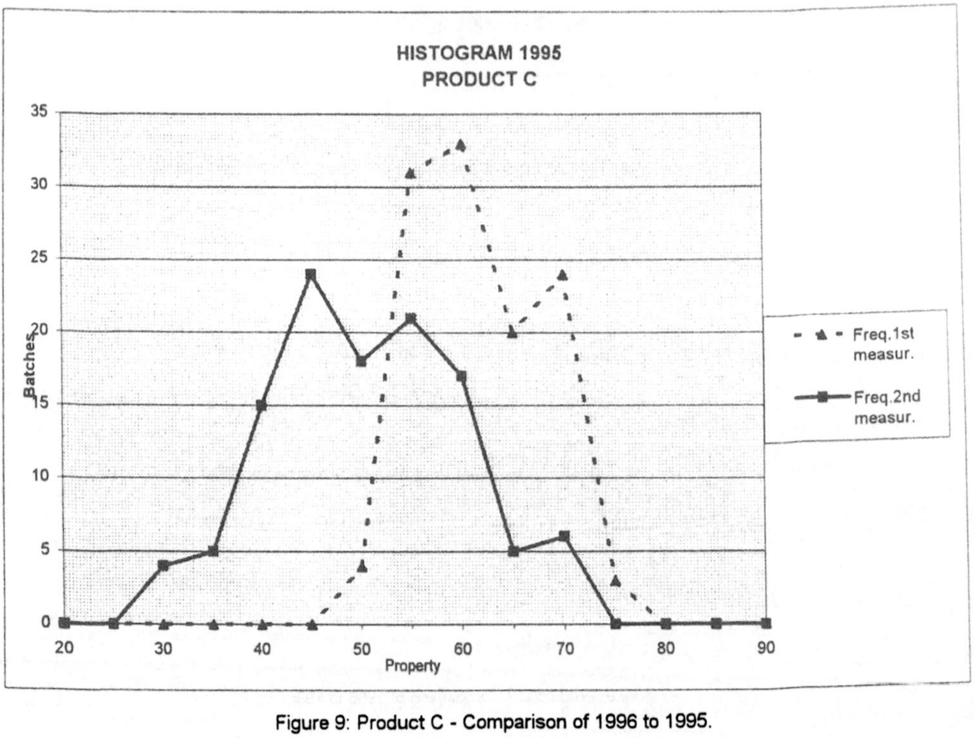

Figure 9: Product C - Comparison of 1996 to 1995.

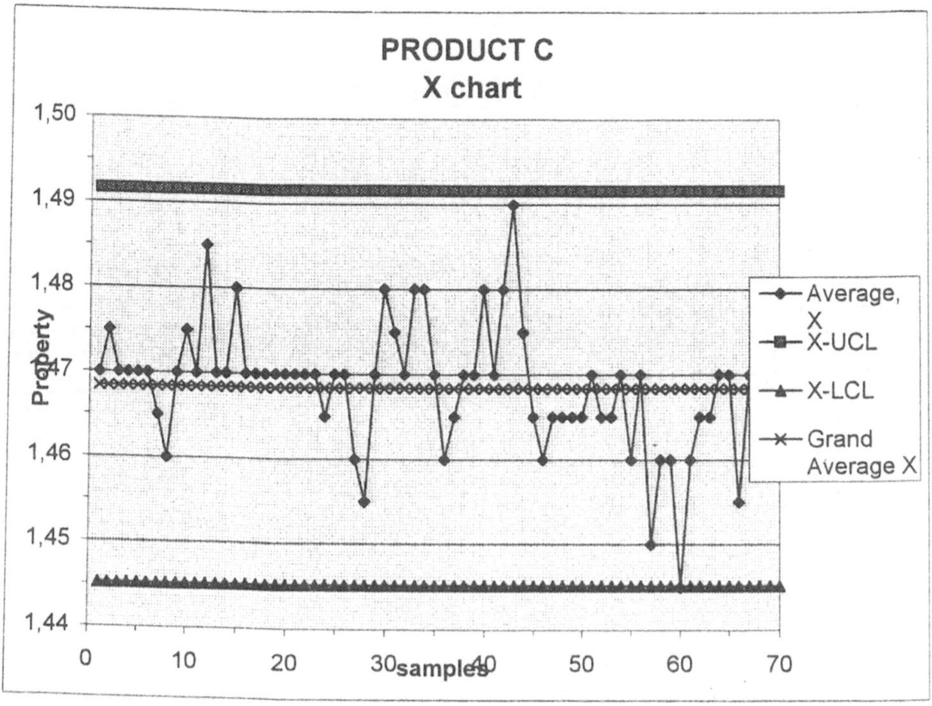

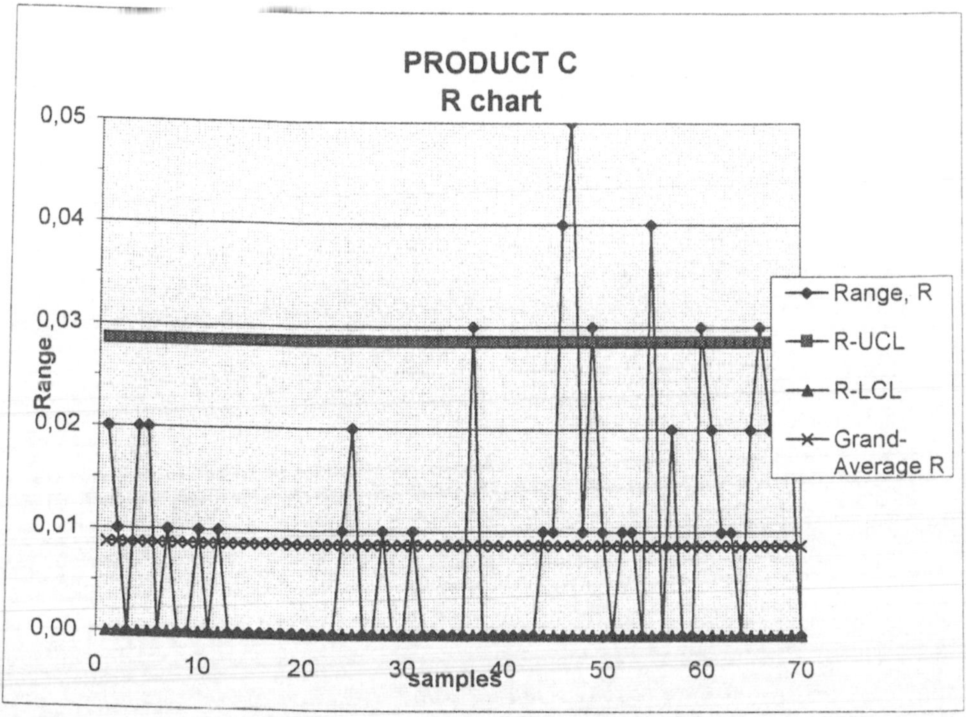

Figure 10: Product C - Control charts

Taking Multivariate Statistics out into Industry - Multivariate Statistical Progress Control and Process Performance Monitoring

ELAINE B. MARTIN and A. JULIAN MORRIS

Centre for Process Analysis, Chemometrics and Control
University of Newcastle upon Tyne,
Newcastle upon Tyne, NE1 7RU
e-mail:- e.b.martin@ncl.ac.uk

Abstract:
The paper describes how the multivariate statistical techniques of principal components analysis (PCA) and projection to latent structures (PLS) can contribute to comprehensive improvements in an industrial process. Univariate SPC has been widely applied in industry but it effectively only detects disturbances related to individual measurement sources (usually quality) and consequently provides little information about the interactions between all the variables which are so important in complex manufacturing processes. Multivariate statistical process control, based upon either PCA or PLS, can provide a diagnostic tool for the on-line monitoring of manufacturing processes and the on-line detection of process malfunctions. A real industrial application is used to highlight the importance of the application of multivariate statistical techniques in industry today for process performance monitoring.

1. Introduction

Major technological challenges currently facing the processing industries include the need to make greater use of existing plant assets, improvements to product yield in line with market perceptions, a reduction in feedstock costs, the minimisation of environmental pollution and the elimination of re-work. This is particularly important in highly demanding situations where processes are subject to varying raw material properties, changing market needs and specifications and fluctuating operating conditions due to equipment or process degradation. The industrial necessity for techniques which enhance process performance requires new methodologies to be adopted which can be used across a spectrum of industrial processing unit operations and on a wide

*Kitsos, C.P. and Edler, L., Eds.:Industrial Statistics, Physica-Verlag, Heidelberg, 1997

range of products. Particularly exciting are the multivariate statistical techniques of principal components analysis and projection to latent structures which are appropriate to a diverse range of industrial problems. Improvements to both the quality and consistency of existing production can be made through the application of these techniques. It is not simply a case of introducing more automation, but of implementing new techniques for data interpretation and process performance monitoring at all levels, from process operators, production engineers and plant managers through to business managers and marketing personnel.

2. Strategic Importance of the Approach

In the early years of a products life, novel products and methods of production may have the advantage of patent protection with the key to competitive advantage being that of intensive research and development and short lead times to market. As the life times of these products evolve many of them become commodities. The management of process capability then becomes an even more complex issue with the primary need being that of the maximisation of production. This necessitates the optimisation of quality, costs, energy and safety, whilst still remaining responsive to changing market conditions. The key to product survival is therefore efficiency and flexibility of production, consistently high quality (zero-defects) production and the maximisation of the use of existing plant assets.

The close monitoring of the operational performance of process plants, their associated instrumentation and control is now seen by industry to be of increasing strategic importance. Multivariate Statistical Process Control (MSPC) can provide a diagnostic tool for the comprehensive on-line monitoring of manufacturing processes and the on-line detection of process malfunctions. The benefits to be seen include increased plant flexibility, improved product consistency and the early warning of process faults and malfunctions.

3. An Overview of Statistical Process Control

Statistical Process Control (SPC) has formed the basis of traditional process performance monitoring and the detection of process malfunctions for the past 80 years. The objective of SPC is to monitor the performance of a process over time to verify that it remains in a state-of-statistical-control. Univariate statistical process control charts, such as Shewhart charts, ($\overline{X}$ and Range charts), CUSUM (cumulative sum) plots, and EWMA (exponentially weighted moving average) charts are typically used for the monitoring of a small number of quality variables. These charts compare current process performance against process behaviour when the product being produced was known to conform to pre-assigned specification limits or when the process continued to operate within statistically derived control limits. The only variation present being as a result of common cause variation which cannot be eliminated from the process. That is, the process is said to be in a state of 'statistical control'.

Univariate statistical methods address the analysis of individual variables. Univariate control charts are based upon the 'magnitude' of the variable deviations and any 'directional' information resulting from variable interactions is ignored. It is therefore not surprising that this can lead to misleading information being presented to operators as to when the process is in a state of statistical control since the interdependence between the variables is ignored. This can have serious consequences since the operational failure of manufacturing plants and their associated instrumentation, controllers and control manipulators is of considerable importance in the management of a process. Process malfunctions can lead to reduced product quality, reduced production, plant shutdowns, increased re-working, increased environmental impact and a low return on plant assets. The need for early warning of potential production problems is therefore a key issue.

4. Multivariate Statistical Process Control

Multivariate statistical process control methods (MSPC) address some of the limitations of univariate monitoring techniques by considering all the data simultaneously and extracting information on the 'directionality' of the process variations. That is, the behaviour of one variable relative to the others. The basis of multivariate statistical process control (MSPC) are the statistical projection techniques of Principal Components Analysis (PCA), Jolliffe (1986) and Projection to Latent Structures (PLS), Höskuldsson, (1988).

4.1 Principal Components Analysis

PCA seeks to find a few linear combinations which can be used to summarise the data with a minimal loss of information. This reduction in dimensionality can be described as a 'parsimonious summarisation' of the data. Let $X = \mathbf{x}_1, \mathbf{x}_2, \mathbf{x}_3, \ldots\ldots, \mathbf{x}_m$ be an m-dimensional data set describing the process variables. The first principal component is that linear combination of the columns of X which describe the greatest amount of variability in X, $t_1 = Xp_1$ subject to $|p_1| = 1$. In m-dimensional space, p_1 defines the direction of greatest variability, and t_1 represents the projection of each object onto p_1. The second principal component is that linear combination defined by $t_2 = E_1 p_2$ which has the next greatest variance subject to $|p_2| = 1$ and subject to the condition that it is orthogonal to the first principal component, t_1 :-

$$t_2 = E_1 p_2 \text{ where } E_1 = (X - t_1 p_1^T)$$

This procedure is essentially repeated until m principal components are calculated. In effect, PCA decomposes the observation vector, X, as

$$X = TP^T = \sum_{i=1}^{m} t_i p_i^T$$

where p_i is an eigenvector of the covariance matrix of X. P is defined as the principal component loading matrix and T is defined to be the matrix of principal component scores. The loadings provide information as to which variables contribute the most to individual principal components, i.e. they are the coefficients in the principal components model;.whilst information on the clustering of the samples and the identification of transitions between different operating regimes is obtained from the scores.

One of the features of PCA is that the less important components often describe the noise in the data. If the process variables are collinear, k principal components (k≤m) will explain the majority of the variability, i.e. a smaller number of principal components than original variables are required to explain the variability in the data. Consequently it is desirable to exclude these components.

$$X = TP^T + E = \sum_{i=1}^{k} t_i p_i^T + E$$

The number of principal components that provide an adequate description of the data can be assessed using a number of techniques. Typically, cross-validation is employed, Wold (1978).

4.2 Prediction of Product Quality

Economic pressures on market perceptions of product quality and reliability has necessitated that the more slowly monitored quality measures be predicted from the more rapidly monitored process variables - inferential estimation. This can be achieved through the application of regression methods, i.e. the identification of a relationship between the quality information and the process variables, hence knowledge of the final quality of the product may be predicted prior to that obtained from the quality control laboratory, for some processes this can result in a saving of many hours. The most widely applied technique is multiple linear regression (MLR). However this approach is less appropriate in multi-dimensional systems where variables are highly correlated, since the model coefficients can be numerically unstable, small perturbations in the data can result in major changes to the model. Alternative regression approaches such as ridge regression and regularisation methods resolve the problem of singularity but do not automatically reduce the dimensionality of the problem - this is of core importance in process performance monitoring in order not to overload process operators with a large number of charts. The technique of projection to latent structures (PLS) utilises the information on both the process and quality variables and treats both sets of information as being dependent.

In Projection to latent Structures the dimensionality of the X and Y spaces are simultaneously reduced to find the latent vectors for the X and Y spaces which are most highly correlated, i.e. those that not only explain the variation in X, but that variation in X which is most predictive of Y. Given a set of information on m process variables, X, and k quality variables, a factor from the X and Y data are evaluated, t_1 and u_1,

$$t_1 = Xw_1 \text{ and } u_1 = Yc_1$$

These equations are referred to as the outer relation for the X block and Y block, respectively. The vectors w_1 and c_1 are called factor weights. PLS finds the factor weights in such a way that t_1 and u_1, are most correlated to one another. A linear regression is then performed between the first pair of factors t_1 and u_1,

$$u_1 = b_1 t_1 + \varepsilon$$

This relationship is termed the inner relation of the X and Y blocks. The final stage of the algorithm is the regression of X on its factor t_1 and Y on its factor u_1:

$$X = t_1 p_1^T + E \text{ and } Y = u_1 q_1^T + F$$

where E and F are the residuals. The above three steps are repeated with the residuals E and F replacing X and Y, respectively. Up to m pairs of factors can be derived, but in a similar way to PCA, only the first few contain the important information relevant to the operation of the process.

5. Process Performance Monitoring Charts

The primary requirement for the development of multivariate SPC charts is the acquisition of data that is representative of nominal process operation, that is when the plant is producing 'within specification product'. The data population can be obtained from either historical data bases or designed experiments. A representation based on historical data collected when only common cause variation was present, can be constructed using either PCA or PLS. Future behaviour is then referenced against this 'nominal' or 'in-control' representation. The basis of the success of this approach is the recognition that many of the measurements are highly correlated and thus different combinations of the variables may define the same underlying disturbances or events occurring on the process. Consequently, it can be assumed that when the process is producing product within predefined specification limits, the dimensionality of the process can be substantially reduced to a few latent variables (Kresta et al, 1991; MacGregor et al, 1994). Typically, this information is presented graphically in terms of time series plots, and two and three dimensional representations of the new latent variables. The four most common forms of monitoring charts are those of the scores against time, two and three dimensional plots of the scores, Hotelling's T^2 (Hotelling, 1947) and the squared prediction error, SPE.

5.1 Squared Prediction Error

Once a model has been developed from the nominal data set using k principal components or latent variables, the fitted values, can be calculated for each new multivariate observation. These values are then used to evaluate the squared prediction error, SPE, or Q-Statistic for each new observation, i.e. the squared difference between the observed values and the predicted values from the nominal or reference model. The squared prediction error plot provides the user with a facility to identify a previously unidentified event.

5.2 Confidence Bounds

Adopting an approach similar to that for univariate statistical process control, nominal operating regions, i.e. warning and action limits, can be defined for each principal components scores plot based upon standard statistical distributional theory. One approach is to assume that the underlying k-dimensional process is normally distributed. It is possible to determine whether the process is in control by calculating Hotelling's (1947) squared distance for each pair of principal components of interest:

$$T^2 = n(\mu_0 - \bar{x})^T S^{-1}(\mu_0 - \bar{x})$$

T^2 is distributed as the statistic $(n-1)kF/(n-k)$, where F has a central F-distribution with k, n-k degrees of freedom. This relationship is used to establish control limits where there is an 100α% chance of a false alarm, where α is typically of the value 0.05 or 0.01. An out of control signal is identified if:-

$$T_0^2 > \frac{(n-1)kF_{k,n-k,\alpha}}{n(n-k)}$$

Confidence bounds for a scores monitoring chart are comparatively straightforward to evaluate. Rules similar to those adopted for univariate SPC charting methodology for identifying when a process is out of control, or moving out of control, can be applied : (i) two points lie outside the warning limits (i.e. $\alpha = 0.05$), (ii) one point lies outside the action limits (i.e. $\alpha = 0.01$), (iii) seven points consistently increase or (iv) seven points consistently decrease. Adopting a similar approach nominal operating regions can be defined for the squared prediction error plot, Jackson (1979).

5.3 Interpreting the 'Out of Control' Signal and 'Causation Variables' Selection

When a process is recognised to be out-of-control or moving out-of-control, operational personnel need a tool to identify the variables, or combination of variables, that are indicative of a drift in the process operating conditions. Adjustments can then be made to the process to avoid the continued manufacture of non-conforming product. Typically, once the process has been identified as not being in a state of statistical control, it is usually the responsibility of the process operators or process engineers to diagnose the assignable cause(s) and implement the appropriate corrective action. This stage could more easily be carried out through the interrogation of the multivariate control charts constructed using either PCA or PLS by reverting back to the original process variables and examining their contribution to the calculated scores and the squared prediction error. It may be possible to identify the most likely set of original variables whose contribution has increased over that defined from the nominal model and which are possibly reflective of non-conforming behaviour. One possible graphical approach is that of the variable contribution plot, Miller (1993), which describes the change in the new observation variables relative to the average value calculated from the PCA/PLS model.

6. An Industrial Application of PCA and PLS Based Process Performance Monitoring

The implementation of both PCA and PLS based multivariate statistical performance monitoring is illustrated on a large scale industrial chemical manufacturing process. The methodology has already been implemented on-line for this particular process. Details of the process are withheld for reasons of commercial confidentiality.

Thirty six process variables and nine product quality variables were measured on-line, the former at ten minute intervals and the latter with a frequency of thirty minutes. A data-set was initially extracted from the historical database which was believed to be representative of current operating performance. The first stage in the analysis was to eliminate periods of data collected during abnormal process operation. This was achieved through discussions with plant personnel and the analysis of the quality and process information. Of particular interest were abnormal events, plant shutdowns, product quality below desired level, etc. This data was removed and the remaining data set formed the nominal data set. The data monitored during major disturbances and plant shutdowns was retained in a separate data base to investigate the potential of MSPC technology to identify the onset of those process problems. In this way a second 'test' data set comprising three hundred observations was created to allow the validation of the techniques off-line prior to full on-line application.

A principal component analysis was first carried out on the nominal data set. Using cross-validation, six components were identified as the 'optimal' number to explain the majority of the variability in the process variables. Multivariate monitoring charts were then constructed based upon the principal component scores vectors, PC1 and PC2, PC3 and PC4, PC5 and PC6 together with the squared prediction error (SPE) and Hotelling's T^2, Figures 1, 3, 4, 5 and 6 respectively.

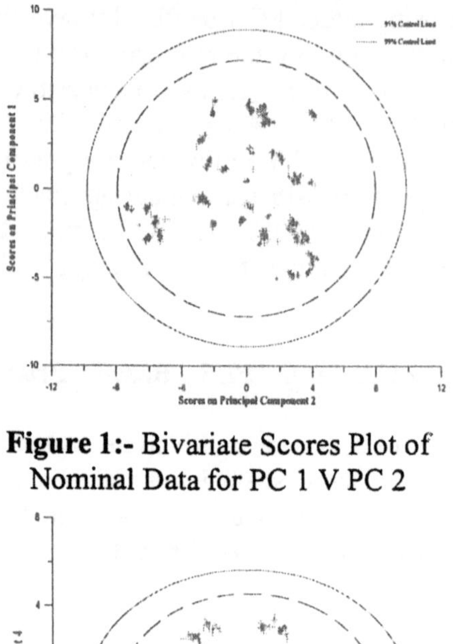

Figure 1:- Bivariate Scores Plot of Nominal Data for PC 1 V PC 2

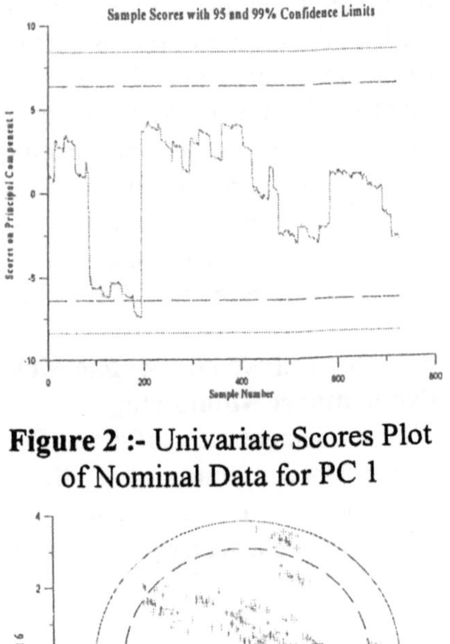

Figure 2 :- Univariate Scores Plot of Nominal Data for PC 1

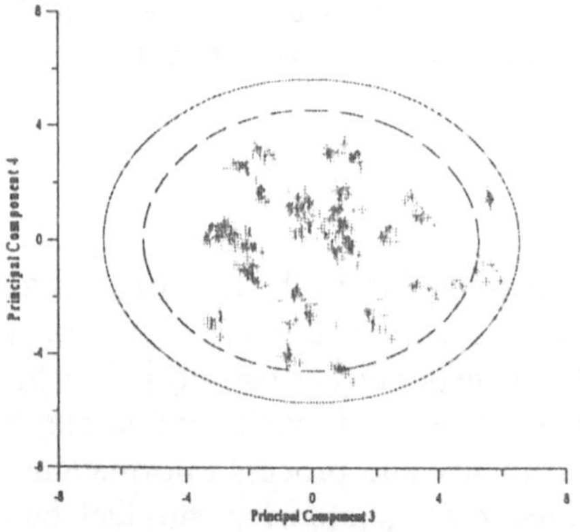

Figure 3 :- Bivariate Scores Plot of Nominal Data for PC 3 V PC 4

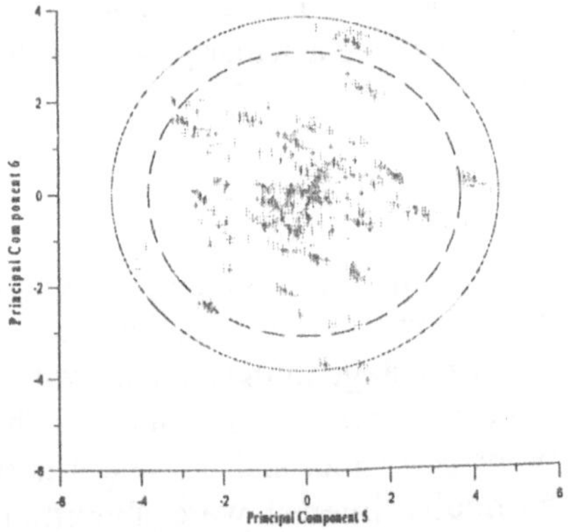

Figure 4:- Bivariate Scores Plot of Nominal Data for PC 5 V PC 6

From the bivariate plot, Figure 1, and the univariate plot for the scores of principal component 1, Figure 2, the process is seen to be in a state of 'statistical control' and subject only to common cause variation. However looking at the lower order principal components, the SPE and Hotelling's T^2, Figures 3, 4 and 5 respectively, it is clear that non-conforming behaviour can be identified. By definition, a proportion of the data will lie outside the confidence bounds, e.g. for a sample of size 20, one sample would lie outside the 95% confidence bound, if the data arose from a normal distribution. In a novel approach to the generation of confidence bounds, the M^2 statistic (Martin and Morris, 1996), flexibility is built-in which enables the user to select the proportion of data which lies inside/outside the bounds. In practice if all the data is from valid production then maybe the bounds should enclose all the nominal data.

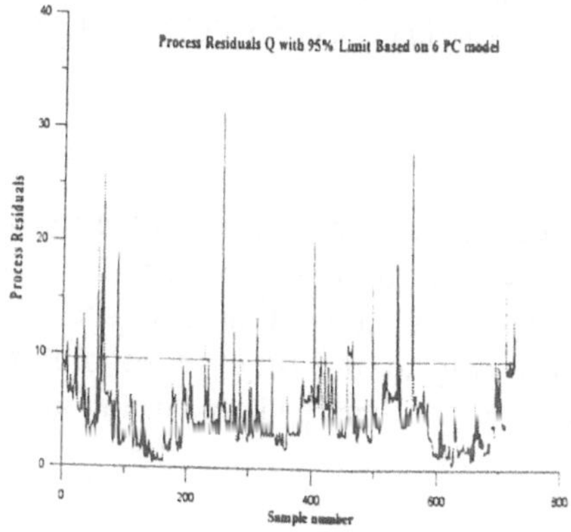

Figure 5 :- SPE for Nominal Data Set

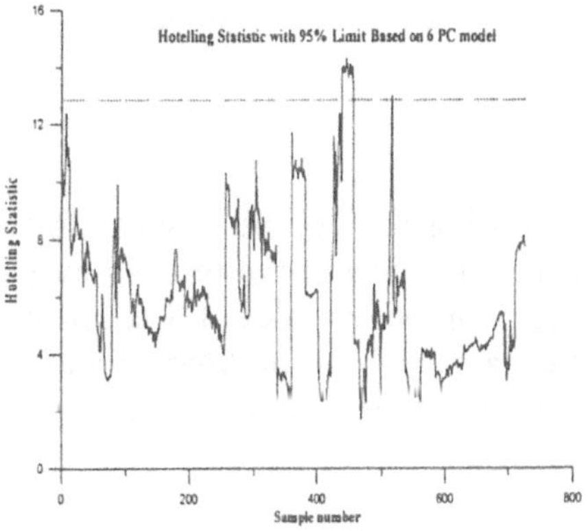

Figure 6 :- Hotellings T^2 for Nominal Data Set

The next stage in the analysis was to reintroduce the data which had previously been identified as 'undesirable,' by the plant engineers, data set 2. Figures 7 and 8 illustrate the impact on both the Hotellings T^2 plot and the bivariate scores plot for principal components 1 and 2. It is clear from both these plots that the non-conforming data would have been identified if process performance monitoring charts had been implemented on the process at the time when the problems had occurred.

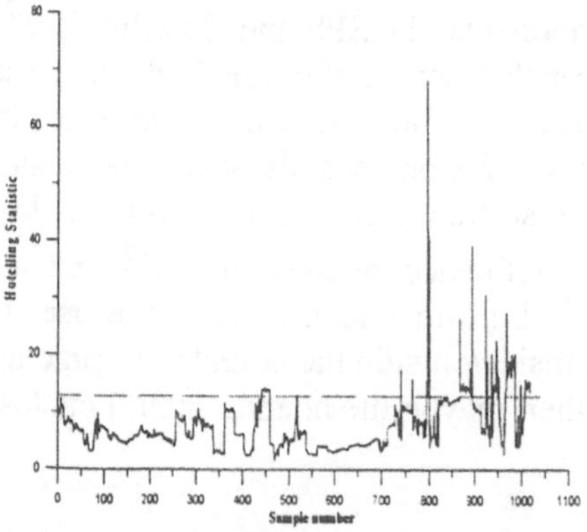

Figure 7 :- Hotellings T^2 for Nominal and 'New' Data Set

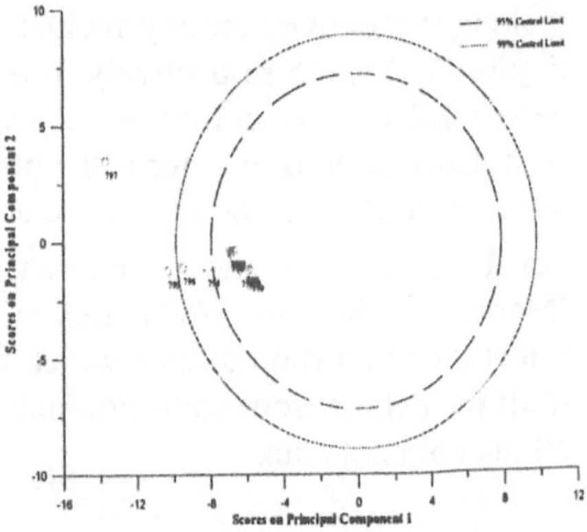

Figure 8:- Bivariate Scores Plot of Nominal Data for PC 1 V PC 2

Reverting back to the original 36 variables, the next step was to identify the apparent cause of the process moving away from desired production. From the contribution plot it was clear that variables three, four and six were indicative of the problem, Figure 9. Looking at the time series plot for variable three, Figure 10, it is clear that a problem had occurred on the process.

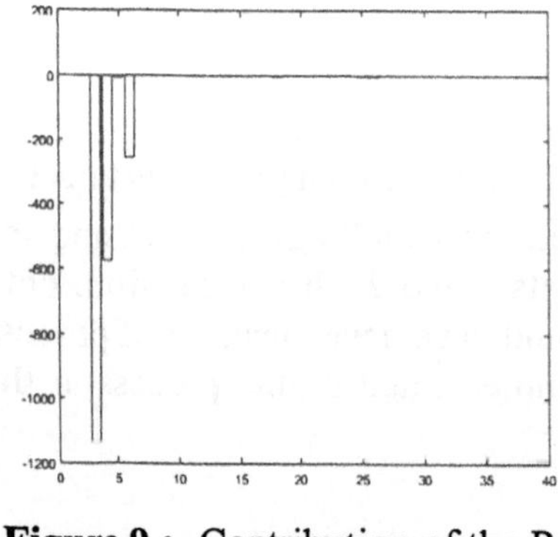

Figure 9 :- Contribution of the Process Variables to the Change in the Scores for Principal Component 1

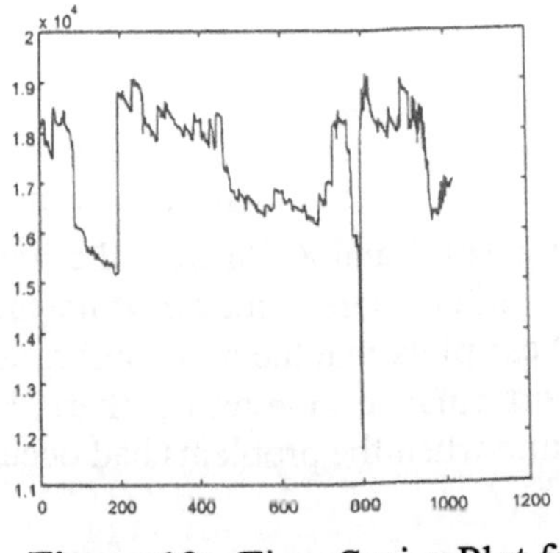

Figure 10:- Time Series Plot for Process Variable 3

Although from the retrospective analysis the identifiable source of the problem was clear, the question was raised, would the problem would have been resolved more rapidly if MSPC had been on-line at that point in time. Under current practice, the plant operator would have been required to have been monitoring either variable three, four or six of the thirty six variables, at that particular point in time when the problem arose. Retrospective analyses such as this has given the operators and engineers confidence to progress the methodology and implement it on-line.

The next phase was to identify a projection to latent structures model which would enable PLS based monitoring to be evaluated on the plant. The results for one of the quality variables is given in Figure 11. Using cross-validation, six latent variables were identified as producing a satisfactory model. This is similar to that for PCA. Figure 12 shows the residuals for the final model which appear to exhibit white noise.

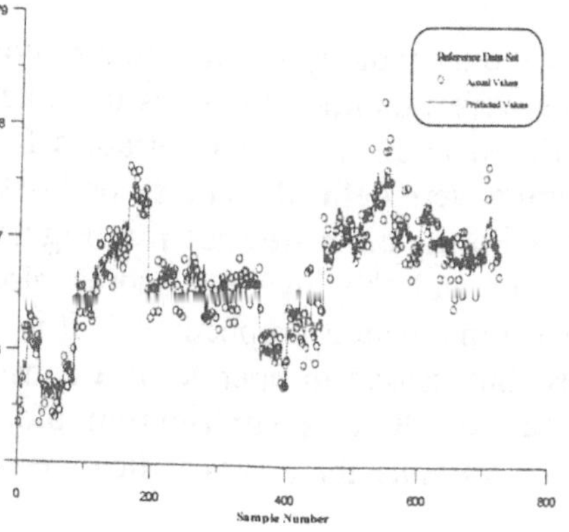

Figure 11 :- Nominal Predictions for Quality Variable

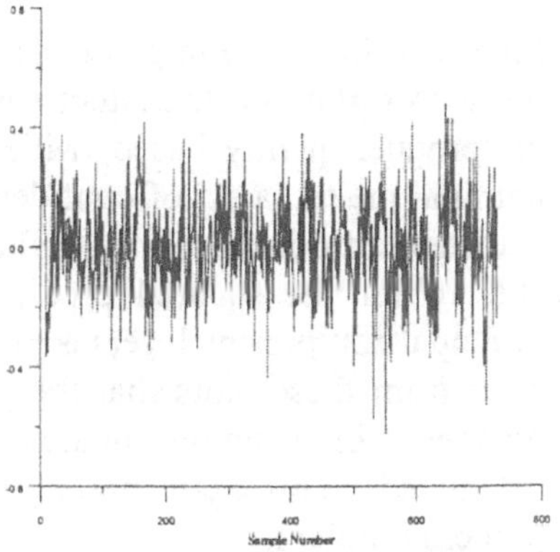

Figure 12 :- Residuals for Quality Variable

The above model was accepted as the basis for the on-line monitoring charts. However, initially it was believed that the process exhibited non-linear behaviour and would require non-linear methodologies such as neural networks to model the process. As a result of extracting a data set which was indicative of desired nominal behaviour, the process was satisfactorily modelled using linear techniques. It appears that it is only the severe non-conforming operation which drives the process into non-linear regions. The linear methods used were sufficient to provide early warning of process malfunctions. Applying the model to the second data set, it was again clearly apparent that early warning of non-conforming behaviour would have resulted as indicated in Figures 13 and 14.

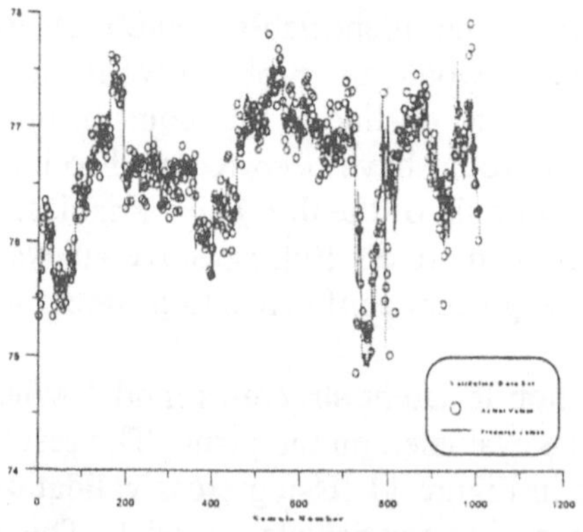

Figure 13 :- Future Predictions for Quality Variable

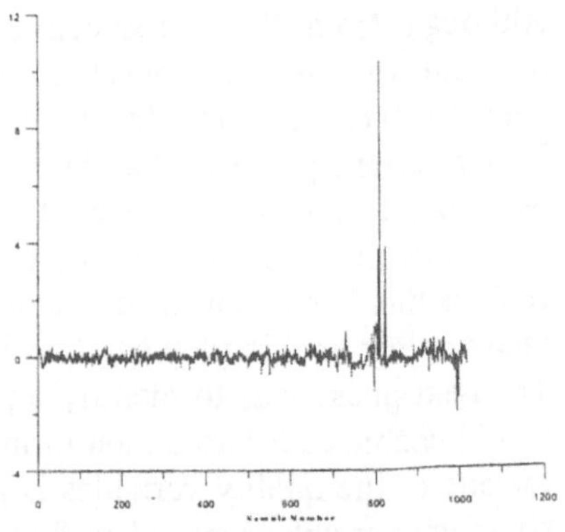

Figure 14 :- Residuals for Quality Variable

Finally, prior to the implementation of the technology for on-line monitoring, the scores plots for the latent variables were examined to assess the potential of 'product quality based' MSPC. Based upon the latent variables for the nominal model, a set of confidence limits were defined. The second data set was then introduced. The resultant monitoring plots are shown in Figures 15 and 16 for the squared prediction error and the bivariate scores plot for principal component 1 versus principal component 2, respectively. It is again clear from these plots that the process has ceased to operate in a consistent manner. In a similar manner to that for PCA, a contribution plot was calculated. This again highlighted that variable three was indicative of the process problem.

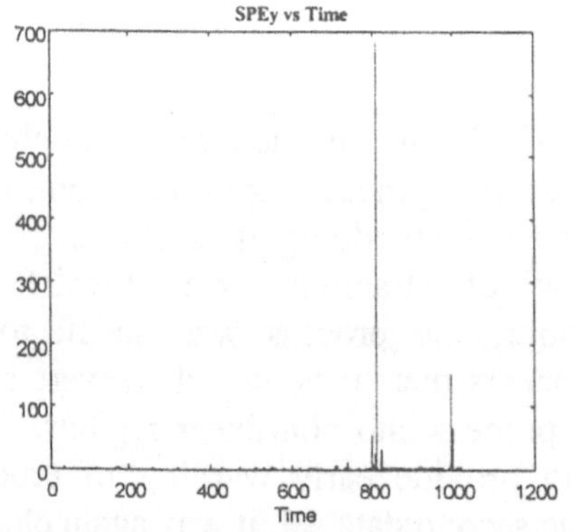

Figure 15 :- SPE plot for PLS based process performance monitoring

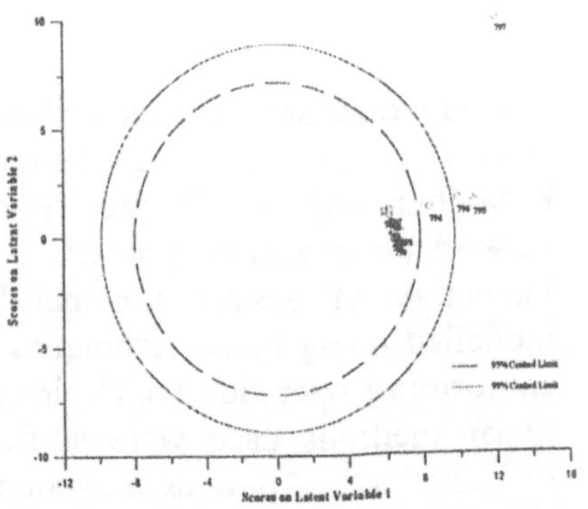

Figure 16 :- Scores plot for PC1 V PC 2

7. Conclusions

The paper has provided an overview of linear multivariate statistical process control based on the statistical projection techniques of principal components analysis and projection to latent structures. The methods proposed take full advantage of the multivariate nature of the data and as a result are more likely to provide early warning of process faults than if univariate SPC techniques, which consider one variable at a time, had been applied. The two areas of particular focus were the identification of the process moving away from the nominal operating region using either univariate or bivariate plots of the scores, squared prediction error plots or Hotelling's T^2, and the identification of the combination of variables reflective of the process moving out of control, through contribution plots. Of particular interest is the necessity to study the lower order principal components to obtain early warning of process malfunctions as opposed to the first one or two components which explain the maximum amount of variability in the data.

There still remain a number of areas of research which require to be addressed before these new approaches to monitoring can be widely adopted by industry in an 'off-the-shelf' manner.

- The further development of density-defined confidence bounds using the M^2 statistic
- The extension of multivariate statistical process control (MSPC) techniques to complete process plants using multi-block projection methods.
- The effect of plant dynamics and plant control loops on the MSPC monitoring methodology and their potential to mask non-conforming process behaviour.
- The sensitivity of the PCA and PLS models.
- The assessment of the need for the development of non-linear multivariate statistical process control.

Within this study the question, 'Does PLS based process performance monitoring give us anything more than PCA based process performance monitoring' was raised. This preliminary study indicated that a similar set of conclusions could be drawn based upon either approach on the condition that the right combination of variables are being monitored and suitable instrumentation and data acquisition systems are in place. Work is on-going to validate the potential of the methodology and its ease of use by operational personnel.

Acknowledgements
The authors acknowledge the support and commitment of the members of the EU ESPRIT Project Process Diagnostics for Plant Performance Enhancement (PROGNOSIS No. 22281) and the European Union for their funding contributions. The academic authors thank their colleagues, in particular Alexandros Simoglou, in the Centre for Process Analysis Chemometrics and Control, University of Newcastle, for their contributions and lively discussions.

References

1. Höskuldsson, A. (1988), "PLS Regression Methods". J. Chemometrics, vol 2, pp 211-228.
2. Hotelling, H., (1947), "Multivariate Quality Control", in Eisenhart, Hastay and Wallis (Eds) *Techniques of Statistical Analysis,* (New York).
3. Jackson, J. E. and Mudholkar, G. S., (1979), "Control Procedures for Residuals Associated with Principal Components Analysis", *Technometrics*, **Vol 21**, No. 3.
4. Jolliffe, I.T. (1986). "Principal Component Analysis". Springer-Verlag, New York.
5. Miller, P., Swanson, R. E., & Heckler, C. F., (1993), "Contribution Plots: The Missing Link in Multivariate Quality Control", *37th Annual Fall Conf. ASQC,* Rochester NY.
6. Kresta, J., MacGregor, J.F., and Marlin, T.E., (1991) "Multivariate Statistical Monitoring of Process Operating Performance", Canadian Journal Chemical Engineering. 68, No. 1.
7. MacGregor, J., (1994a), "Statistical Process Control of Multivariate Processes", *Preprints of the IFAC ADCHEM'94 Conference on Advanced Control of Chemical Processes*, May, Kyoto Japan.
8. Martin, E. B. and Morris, A. J., (1996), "Non-parametric Confidence Bounds for Process Monitoring Charts", *J. of Proc. Cont.*, **Vol 6**, No. 6, 349-358.
9. Wold. S., "Cross-Validatory estimation of the Number of Components in Factor and Principal Components Models". *Technometrics*, **20**, 397-405, (1978)

Robust Estimation of Variance Components with High Breakdown Point in the 1-Way Random Effect Model

STEFFEN UHLIG
Free University Berlin, Institute of Statistics and Econometrics
Garystr. 21 , D-14195 Berlin, Germany
e-mail: steffen_uhlig@ccmailer.wiwiss.fu-berlin.de

Key Words: variance component model, robust estimation, interlaboratory test, breakdown point

Abstract

For the 1-way random effect model three new robust variance component estimators are proposed. Consistency and breakdown properties are derived. The use of the estimators for interlaboratory tests is demonstrated in a practical example.

1 Introduction

Consider the 1-way random effect model

$$Y_{ni} = \mu + \alpha_n + \epsilon_{ni},$$

where $n = 1, \ldots, N$ denotes the factor level and $i = 1, \ldots, I_n$ denotes the single measurement. The random variables $\alpha_1, \ldots, \alpha_N, \epsilon_{11}, \ldots, \epsilon_{NI_N}$ are stochastically independent and normally distributed with expectation zero and variances $Var[\alpha_n] = \sigma_A^2$ and $Var[\epsilon_{ni}] = \sigma_0^2$.

This model can be applied for interlaboratory comparison tests: In order to assess the performance of individual analytical laboratories, several laboratories $n = 1, \ldots, N$ analyse samples of a homogenous test material and obtain measurement values Y_{ni}. However, the use of the classical 1-way random effect model means that all laboratories have the same performance level: This is

*Kitsos, C.P. and Edler, L., Eds.:Industrial Statistics, Physica-Verlag, Heidelberg, 1997

not consistent with the idea of assessing performance levels of the individual laboratories. If there are just a few 'outlier' laboratories, the data of these laboratories can be eliminated by outlier tests (as described in ISO-5725). In interlaboratory comparison tests employed for the certification and licencing of laboratories, there is quite often a large number of laboratories which do not meet the quality requirements. Then it is hardly reasonable to group all laboratories under one population without drawing any distinctions. Rather the extent of the laboratory's measurement deviations in the model should be depending on the performance level of the individual laboratory. Introducing laboratory-specific relative variability coefficients $\gamma_n \geq 1$ and $\delta_n \geq 1$, the statistical model could be extended into a model

$$Y_{ni} = \mu + \delta_n \alpha_n + \gamma_n \epsilon_{ni}.$$

Higher values of δ_n and γ_n correspond with a lower analytical performance, smaller values with a higher performance. $\delta_n = \gamma_n = 1$ corresponds with maximum performance of the laboratory. Maximum performance will be achieved if the technical, organisational and staff-related conditions prevalent in the laboratory meet the given quality requirements. Then the variability of measurement results is caused by the intrinsic variability of the analytical method prescribed.

Note that the parameters of the extended model are not identifiable; additional restrictions are necessary. In this paper it will be assumed that more than 50% of the laboratories meet the quality requirements, i.e.

$$\#\{n = 1, \ldots, N \mid \delta_n = \gamma_n = 1\} > 0.5 \cdot N$$

Therefore we can reconsider the old 1-way random effect model

$$Y_{ni} = \mu + \alpha_n + \epsilon_{ni},$$

but we have to take into account that possibly up to <50% of the laboratories are 'outliers'.

In this paper robust variance component estimators for σ_0^2 and $\sigma_Y^2 = \sigma_0^2 + \sigma_A^2$ are presented. These estimators are based on generalized L-estimators (see Serfling 1984 and Rousseeuw and Croux 1993a) and have approximately 50% breakdown point.

2 Definitions

2.1 Breakdown point

The breakdown point of an estimator is the smallest proportion of observations replaced by outliers that can cause a 'breakdown' of the estimator. Several breakdown point definitions are available. For (location) estimators in designed experiments see Mueller (1993) and for scale estimators in the univariate case see Rousseeuw and Croux (1993b). However, these concepts cannot be applied directly for the 1-way random effect model, because in the latter two sources of error, factors and replicates, have to be taken into account. Associated with the factor levels is the following definition. It is a replacement version of the breakdown point based on the random variables $y_n = (y_{n1}, \dots, y_{nI_n})$, $n = 1, \dots, N$:

Definition 1 *The first order breakdown point of the scale estimator $S = S(y_1, \dots, y_N)$ for the error variance σ_0^2 or for the total variance σ_Y^2 is defined by*

$$\begin{aligned} e^*_{N,1}(S,(y_1,\dots,y_N)) = \frac{1}{N}\min\{M| \\ \sup\{S(z_1,\dots,z_N)=\infty|\,(z_1,\dots,z_N)\in L_{N,M,1}(y_1,\dots,y_N)\}\ \textit{or} \\ \inf\{S(z_1,\dots,z_N)=0|\,(z_1,\dots,z_N)\in L_{N,M,1}(y_1,\dots,y_N)\}\} \end{aligned}$$

where $L^{(1)}_{N,M}(y_1, \dots, y_N)$ denotes the set of all samples that are obtained by replacing M data vectors $y_{k_1}, \dots, y_{k_M}$ by arbitrary vectors of same size, i.e.

$$\begin{aligned} L^{(1)}_{N,M}(y_1,\dots,y_N) &= \{(z_1,\dots,z_N)| \\ &\quad \#\{n=1,\dots,N|\,z_n \neq y_n\} \leq M\}\} \end{aligned}$$

The first order breakdown point measures the smallest proportion of contaminated factor levels that can cause the estimated value tending to ∞ or to 0.

Associated with the observations itselves is the second order breakdown point:

Definition 2 *The second order breakdown point of the scale estimator $S = S(y_1, \dots, y_N) = S(y_{ni})$ is defined by*

$$\begin{aligned} e^*_{N,2}(S,(y_1,\dots,y_N)) = \frac{1}{K}\min\{M| \\ \sup\{S(z_1,\dots,z_N)=\infty|\,(z_1,\dots,z_N)\in L^{(2)}_{I_1,\dots,I_N,M}(y_1,\dots,y_N)\}\ \textit{and} \\ \inf\{S(z_1,\dots,z_N)=0|\,(z_1,\dots,z_N)\in L^{(2)}_{I_1,\dots,I_N,M}(y_1,\dots,y_N)\}\} \end{aligned}$$

where $L^{(2)}_{I_1,\dots,I_N,M}(y_1, \dots, y_N)$ denotes the set of all samples that are obtained by replacing M observations y_{ni} by arbitrary variables, i.e.

$$\begin{aligned} L^{(2)}_{I_1,\dots,I_N,M}(y_1,\dots,y_N) = \{(z_1,\dots,z_N)| \\ \#\{(n,i)|\,n=1,\dots,N,\ i=1,\dots,I_n,|\,z_{ni}\neq y_{ni}\} \leq M\}\} \end{aligned}$$

The second order breakdown point measures the smallest proportion of contaminated observations that can cause the estimated value tending to ∞ or to 0.

In balanced models the following unequation holds:

Lemma 1 *If $I_1 = \dots, I_N$, for every estimator S*

$$e^*_{N,1}(S)(y_1,\dots,y_N) \geq e^*_{N,2}(S)(y_1,\dots,y_N).$$

2.2 Estimating σ_0

Assume that $I_n \geq 2$ for all $n = 1, \ldots, N$. For estimating σ_0 the following estimators will be considered:

$$\begin{aligned} S_{N,1} &:= \frac{1}{\sqrt{2}\Phi^{-1}((0.5+1)/2)} H_{N,1}^{-1}(0.5) = 1.048 H_{N,1}^{-1}(0.5) \\ S_{N,2} &:= \frac{1}{2\Phi^{-1}((0.25+1)/2)} H_{N,2}^{-1}(0.25) = 1.569 H_{N,2}^{-1}(0.25), \end{aligned}$$

where

$$\begin{aligned} H_{N,1}(x) &:= \frac{1}{N}\sum_{n=1}^{N} \frac{2}{I_n(I_n-1)} \sum_{1\leq i<j\leq I_n} 1\{|Y_{ni}-Y_{nj}| \leq x\} \\ H_{N,2}(x) &:= \frac{1}{\binom{N}{2}} \sum_{1\leq n<m\leq N} \frac{1}{I_n(I_n-1)I_m(I_m-1)} \sum_{(i,j,k,l)\in T_{n,m}} \\ &\quad 1\{|Y_{ni}-Y_{nj}-Y_{mk}+Y_{ml}| \leq x\} \\ &\quad + 1\{|Y_{ni}-Y_{nj}+Y_{mk}-Y_{ml}| \leq x\} \end{aligned}$$

and

$$T_{n,m} := \{(i,j,k,l) | i,j = 1,\ldots,I_n;\ i<j;\ k,l = 1,\ldots,I_m;\ k<l\}.$$

$H_{N,1}(x)$ can be interpreted as the distribution function of the absolute intra-level differences. Notice that the weight due to the difference $Y_{ni} - Y_{nj}$ is reciprocal to the number of absolute differences of the factor level n. Therefore $H_{N,1}(x)$ can be written

$$\begin{aligned} H_{N,1}(x) &= \frac{1}{N}\sum_{n=1}^{N} \frac{2}{I_n(I_n-1)} \sum_{1\leq i<j\leq I_n} 1\{|Y_{ni}-Y_{nj}| \leq x\} \\ &= \sum_{(i_1,j_1,i_2,j_2,\ldots,i_N,j_N)\in T_{N,1}} \frac{1}{N}\sum_{n=1}^{N} 1\{|Y_{ni_n}-Y_{nj_n}| \leq x\}/M_N \\ &= \sum_{(i_1,j_1,i_2,j_2,\ldots,i_N,j_N)\in T_{N,1}} G^{(1)}_{i_1,j_1,i_2,j_2,\ldots,i_N,j_N}(x)/M_N, \end{aligned}$$

where

$$G_{i_1,j_1,i_2,j_2,\ldots,i_N,j_N}(x) := \frac{1}{N}\sum_{n=1}^{N} 1\{|Y_{ni_n}-Y_{nj_n}| \leq x\}$$

denotes the empirical distribution function of the absolute differences $|Y_{ni_n} - Y_{nj_n}|$, $n = 1, \ldots, N$ and

$$\begin{aligned} T_{N,1} &:= \{(i_1,j_1,i_2,j_2,\ldots,i_N,j_N) | \\ &\quad 1\leq i_1<j_1\leq I_1;\ 1\leq i_2<j_2\leq I_2;\ \ldots;\ 1\leq i_N<j_N\leq I_N\} \end{aligned}$$

and $M_{N,1} := \#(T_{N,1}) = \frac{1}{2^N}\Pi_{n=1}^N I_n(I_n - 1)$. Therefore $H_{N,1}$ can be interpreted as the arithmetic mean of all distribution functions $G_{i_1,j_1,i_2,j_2,\dots,i_N,j_N}(x)$. In the balanced case, $I_1 = \dots = I_N$, the function $H_{N,1}(x)$ equals the distribution function of all absolute within-level differences. In that case the estimator $S_{N,1}$ is simply a Fisher-consistent version of the median of all these differences.

$H_{N,2}(x)$ can be interpreted as the distribution function of the absolute second order differences of each factor level pair. Note that the weight of each difference is reciprocal to the number of the differences associated with the same factor level pair. Hence each pair of factor levels has the same weight in $H_{N,2}(x)$. In the balanced case, $I_1 = \dots = I_N$, the function $H_{N,2}(x)$ equals the distribution function of all absolute inter-level differences of the within-level differences. In that case the estimator $S_{N,2}$ corresponds to the lower quartile of these differences.

Consider the model with two replicates at each factor level, i.e. $I_n = 2$. Then $S_{N,2}$ equals the lower quartile of all absolute differences $|Y_{n1} - Y_{n2} - Y_{m1} + Y_{m2}|$ and $|Y_{n1} - Y_{n2} + Y_{m1} - Y_{m2}|$. This is similar but not equal to the estimator proposed by Lischer (1996)

$$S_{N,Lischer} = Q_n(Y_{11} - Y_{12}, Y_{21} - Y_{22}, \dots, Y_{N1} - Y_{N2}).$$

Here Q_n denotes the scale estimator for univariate samples defined by Rousseeuw and Croux (1993a and 1993b)

$$Q_n = 2.2194\{|y_i - y_j|;\ i < j\}_{(k)},$$

where $k = \binom{h}{2} \approx \binom{n}{2}/4$ and $h = [n/2] + 1$.

Rousseeuw and Croux show that Q_n is consistent for σ under normal distribution. Its asymptotic efficiency under normal distribution is 82.3%. $Q_n(Y_{11} - Y_{12}, Y_{21} - Y_{22}, \dots, Y_{N1} - Y_{N2})$ equals approximately the lower quartile of all absolute differences $|Y_{n1} - Y_{n2} - Y_{m1} + Y_{m2}|$ but takes not into account $|Y_{n1} - Y_{n2} + Y_{m1} - Y_{m2}|$. Therefore the estimator of Lischer is not invariant against permutations of Y_{n1} and Y_{n2}.

2.3 Estimating σ_Y

For estimating σ_Y the following estimator will be considered:

$$S_{N,3} := \frac{1}{\sqrt{2}\Phi^{-1}((0.25+1)/2)} H_{N,3}^{-1}(0.25) = 2.2194 H_{N,3}^{-1}(0.25),$$

where

$$H_{N,3}(x) := \frac{1}{\binom{N}{2}} \sum_{1 \le n < m \le N} \frac{1}{I_n I_m} \sum_{i=1}^{I_n} \sum_{j=1}^{I_m} 1\{|Y_{ni} - Y_{mj}| \le x\}$$

and

$$T_{n,m} := \{(i,j,k,l) | i,j = 1,\dots,I_n;\ i \ne j;\ k,l = 1,\dots,I_m;\ k \ne l\}.$$

Like $H_{N,1}$ and $H_{N,2}$, the function $H_{N,3}(x)$ can be interpreted as the distribution function of the absolute inter-level differences. Note that the weight of each difference is reciprocal to the number of the differences of the associated factor level pair. Hence each pair of factor levels has the same weight in $H_{N,3}(x)$. In the balanced case, $I_1 = \ldots = I_N$, the function $H_{N,3}(x)$ equals the distribution function of all absolute inter-level differences. In that case the estimator $S_{N,3}$ corresponds to the lower quartile of these differences.

3 Breakdown point and consistency

Theorem 1 (First order breakdown point) *(1) At any sample $\{y_{ni}\}$ in which no two points coincide, the estimators $S_{N,1}$ and $S_{N,3}$ have the asymptotic first order breakdown point*

$$\lim_{N\to\infty} e^*_{N,1}(S_{N,1},(y_1,\ldots,y_N)) = \lim_{N\to\infty} e^*_{N,1}(S_{N,3},(y_1,\ldots,y_N)) \quad = \quad 0.5.$$

(2) At any sample $\{y_{ni}\}$ in which all differences $|y_{ni} - y_{nj}|$ are pairwise different and not equal to zero, the estimator $S_{N,2}$ has the asymptotic first order breakdown point

$$\lim_{N\to\infty} e^*_{N,1}(S_{N,2},(y_1,\ldots,y_N)) \quad = \quad 0.5$$

Note that the asymptotic first order breakdown point is not depending on the number of replicates I_n. The second order breakdown point is depending on the number of replicates. The following theorem presents the second order breakdown points for the balanced case with two replicates, i.e. $I_1 = \ldots = I_N = 2$.

Theorem 2 (Second order breakdown point) *Assume that $I_1 = \ldots = I_N = 2$.*

(1) At any sample $\{y_{ni}\}$ in which no two points coincide, the estimator $S_{N,1}$ has second order breakdown point

$$e^*_{N,2}(S_{N,1},(y_1,\ldots,y_N)) \quad = \quad \frac{[(N+1)/2]}{2N}.$$

(2) At any sample $\{y_{ni}\}$ in which all differences $|y_{ni} - y_{nj}|$ are pairwise different and not equal to zero, the estimator $S_{N,2}$ has second order breakdown point

$$e^*_{N,2}(S_{N,2},(y_1,\ldots,y_N)) \quad = \quad \frac{[(N+1)/2]}{2N}.$$

(3) At any sample $\{y_{ni}\}$ in which no two points coincide, the estimator $S_{N,3}$ has second order breakdown point

$$e^*_{N,2}(S_{N,3},(y_1,\ldots,y_N)) \quad = \quad \frac{N-1}{2N}.$$

Appearantly the second order breakdown point of $S_{N,1}$ and $S_{N,2}$ equals approximately 25%, whereas $S_{N,3}$ has approximately 50% second order breakdown point.

Theorem 3 (Consistency) *(1) Let $I_n \geq 2$ for all $n \geq 1$. Then $S_{N,1}$ and $S_{N,2}$ converge in probability to σ_0, i.e. as $N \to \infty$*

$$S_{N,1} \xrightarrow{p} \sigma_0 \text{ and } S_{N,2} \xrightarrow{p} \sigma_0.$$

(2) $S_{N,3}$ converges in probability to σ_Y, i.e. as $N \to \infty$

$$S_{N,3} \xrightarrow{p} \sigma_Y.$$

4 Simulation study and discussion

In order to get insight into the finite sample properties of the robust estimators, a simulation study was performed with 10000 runs per parameter setting. The following tables list the results for the balanced model with $I_n = 2$ replicates. The calculation of the efficiency is based on the standardized variance

$$\text{Std.Var.} = N \cdot Var(S_N)/Mean(S_N)$$

where $Var(S_N)$ and $Mean(S_N)$ denote the values obtained from the simulation study. It was argued by Bickel and Lehmann (1976) that this standardization is needed to obtain a natural measure of accuracy for scale estimators.

Simulation results of $S_{N,1}$

	$N=8$	$N=12$	$N=16$	$N=20$	$N=30$	$N=50$
Std.Var.:	1.041	1.121	1.168	1.222	1.273	1.342
Mean:	1.044	1.026	1,020	1,016	1.013	1.007
Eff.:	49.6%	45.2%	43.5%	41.5%	39.1%	38.4%

Simulation results of $S_{N,2}$

	$N=8$	$N=12$	$N=16$	$N=20$	$N=30$	$N=50$
Std.Var.:	0.721	0.703	0.679	0.670	0.647	0.647
Mean:	1.310	1.202	1.145	1.118	1.079	1.045
Eff.:	71.6%	74.8%	74.8%	75.7%	76.9%	79.6%

Simulation results of $S_{N,Lischer}$

	$N=8$	$N=12$	$N=16$	$N=20$	$N=30$	$N=50$
Std.Var.:	0.943	0.882	0.820	0.787	0.734	0.706
Mean:	1.496	1.32	1.235	1.191	1.127	1.074
Eff.:	54.7%	59.5%	54.6%	64.4%	67.8%	72.9%

Simulation results of $S_{N,3}$

	$N=12$ $\sigma_A=1$ $\sigma_0=0$	$N=12$ $\sigma_A=0$ $\sigma_0=1$	$N=30$ $\sigma_A=1$ $\sigma_0=0$	$N=30$ $\sigma_A=0$ $\sigma_0=1$
Std.Var.:	0.878	0.368	0.741	0.346
Mean:	0.998	0.991	0.996	0.992
Rel.Eff.:	65%	72%	69%	76%

The simulation study performed also included models with 3 and 4 replicates (not listed in the table). In all cases the efficiency of the robust estimators was even better as with $I = 2$ replicates. It can be concluded that $S_{N,2}$ and $S_{N,3}$ are promising estimators having high breakdown point and good finite sample efficiencies. It is interesting that the estimator $S_{N,Lischer}$ is less efficient than $S_{N,2}$, although it is very similar to the latter. $S_{N,1}$ is also less efficient than $S_{N,2}$, but it reveals the advantages of less computational effort and a smaller gross error sensitivity (which has not been considered in this paper). Therefore the final choice between $S_{N,1}$ and $S_{N,2}$ should depend on the practical application.

5 Example

An interlaboratory comparison test was untertaken to assess the performance of $N = 23$ chemical laboratories with respect to the analysis of arsenic in soil. Each of the laboratories analyzed the arsenic content twice. The following table contains the ordered measurement results.

Lab	1	2	3	4	5	6	7	8	9	10	11	12
	0.19	1.60	1.90	2.00	2.00	2.00	2.00	2.00	2.02	2.10	2.10	2.44
	0.21	1.60	2.10	2.50	3.00	1.90	2.00	2.00	2.22	2.00	2.30	2.40

Lab	13	14	15	16	17	18	19	20	21	22	23
	2.50	2.70	2.74	3.80	4.00	4.80	6.20	7.00	8.00	8.98	27.00
	2.40	2.20	2.69	4.00	4.20	5.20	5.80	9.00	7.00	25.83	30.00

Approximately 60% of the laboratories are in a small range between 1.6 and 2.75, whereas some of the other laboratories provided results which are far from this interval. The reason for the considerable heterogenity of the data could be that the laboratories use different analytical methods or have different analytical equipment.

In a classical analysis according to ISO 5725 two outliers (lab 22 and 23) are detected. After elimination of these laboratories one obtains the estimates $\hat{\sigma}_0 = 0.41$ and $\hat{\sigma}_Y = 1.99$. Under the assumption of the normal distribution, the latter fits not very well to the fact, that the results of 60% of the labs are within an interval of length 1.15.

A 'better' fit is provided by the robust method: One obtains the estimate

$\hat{\sigma}_Y = S_{23,3} = 1.11$. For σ_0 one obtains $S_{23,1} = 0.21$ and $S_{23,2} = 0.31$. The latter is higher because of the higher gross error sensitivity of $S_{N,2}$. But both estimates are smaller than the classical estimate according to ISO 5725. Note that the results of ISO 5725 are highly depending on the results of the outlier tests. Elimination of three or four outlier laboratories would provide results which are similar to those of the robust method.

References

Bickel, P.J. and Lehmann, E.L. (1976), Descriptive Statistics for Nonparametric Models III: Dispersion. The Annals of Statistics, 4, 1139-1158.

ISO-5725 (1992), Accuracy (Trueness and Precision) of Test Measurements. International Organisation for Standardisation, Geneva, Switzerland.

Lischer, P. (1996), Robust Statistical Methods in Interlaboratory Analytical Studies. In: Robust Statistics, Data Analysis, and Computer Intensive Methods. Ed. H. Rieder, Springer, 251-264.

Mueller, C.H. (1995), Breakdown points for designed experiments. Journal of Statistical Planning and Inference, 45, 413-427.

Rousseeuw, P.J. and Croux, C. (1993a), Explicit Scale Estimators with High Breakdown Point. University of Antwerp, Department of Mathematics & Computer Science. Report 92-0988.

Rousseeuw, P.J. and Croux, C. (1993b), Alternatives to the Median Absolute Deviation. Journal of the American Statistical Association, 88, 1273-1283.

Serfling, R.J. (1984), Generalized L , M-, and R-Statistics. The Annals of Statistics, 12, 76-86.

Proofs will be published elsewhere.

Sources of Inaccuracy in the Mean and Variance Models for Quality Improvement Problems

IVAN. N. VUCHKOV AND LIDIA. N. BOYADJIEVA

University of Chemical Technology and Metallurgy, 1756 Sofia, Bulgaria

Key Words:

Quality improvement, Taguchi method, model based approach, mean and variance of performance characteristic, errors in parameters, tolerances, manufacturing imperfections, external noises

Abstract:

A model based approach is discussed for quality improvement. The mean and variance models of performance characteristic are given. The influence of the following sources of inaccuracy on the mean and variance is considered: incorrect regression model of performance characteristic, neglecting the high order error distribution moments and random character of the regression coefficient estimates.

1. Introduction

In this paper we consider a model based alternative of well known Taguchi method. It involves development and use of mean and variance models of the performance characteristic in mass production and usage. Models of performance characteristic's mean and variance are proposed by Vuchkov and Boyadjieva (1981, 1988, 1989, 1990a, 1990b, 1992, 1994), Vuchkov, Boyadjieva and Damyanov (1987), Box and Fung (1986), Box and Jones

*Kitsos, C.P. and Edler, L., Eds.:Industrial Statistics, Physica-Verlag, Heidelberg, 1997

(1992), Vinning and Myers (1990). In our approach we consider a performance characteristic η which depends on following factors:

$\mathbf{p} = \left(p_1 \quad p_2 \quad \ldots \quad p_m\right)^T$, product and/or process parameters,

$\mathbf{e} = \left(e_1 \quad e_2 \quad \ldots \quad e_m\right)^T$, errors in the parameters (tolerances, manufacturing imperfections, etc.),

$\mathbf{n} = \left(n_1 \quad n_2 \quad \ldots \quad n_q\right)^T$, external noise factors, (environmental temperature, humidity, vibrations, etc.

The measured value of the performance characteristic in use of the product is

$$y(\mathbf{p}+\mathbf{e},\mathbf{n}) = \eta(\mathbf{p}+\mathbf{e},\mathbf{n}) + \varepsilon,$$

where ε is random output error. We assume that e, n and ε are independent and normally distributed with zero expectations and following covariance matrices:

$$\Sigma_{\mathbf{e}} = diag\left(\sigma_1^2, \sigma_2^2, \ldots, \sigma_m^2\right), \quad \Sigma_{\mathbf{n}} = diag\left(\sigma_{n1}^2, \sigma_{n2}^2, \ldots, \sigma_{nq}^2\right).$$

The variance of the random noise ε is σ_ε^2. Under the assumption for normality of distributions the third order moments of e, n and ε are equal to zero, while the fourth order moments are given as

$$E\left(e_i^4\right) = 3\sigma_i^4,\ E\left(n_j^4\right) = 3\sigma_{nj}^4,\ E\left(\varepsilon^4\right) = 3\sigma_\varepsilon^4,\ i = 1,2,\ldots,m;\ j = 1,2,\ldots q.$$

Suppose that following second order polynomial model of the performance charcteristic is used :

$$y(\mathbf{p},\mathbf{n}) = \beta_0 + \sum_{i=1}^{m} \beta_i p_i + \sum_{i=1}^{m-1} \sum_{j=i+1}^{m} \beta_{ij} p_i p_j + \sum_{i=1}^{m} \beta_{ii} p_i^2 + \sum_{i=1}^{q} \alpha_i n_i + \\ + \sum_{i=1}^{q} \alpha_{ii} n_i^2 + \sum_{i=1}^{q-1} \sum_{j=i+1}^{q} \alpha_{ij} n_i n_j + \sum_{i=1}^{m} \sum_{j=1}^{q} \gamma_{ij} p_i n_j + \varepsilon. \qquad (1)$$

This equation can also be written in matrix form as:

$$y(\mathbf{p},\mathbf{n}) = \beta_0 + \beta^T\mathbf{p} + \mathbf{p}^T\mathcal{B}\mathbf{p} + \alpha^T\mathbf{n} + \mathbf{n}^T\mathcal{A}\mathbf{n} + \mathbf{p}^T\mathcal{G}\mathbf{n} + \varepsilon. \qquad (2)$$

where

$$\tilde{y} = E[y(\mathbf{p}+\mathbf{e},\mathbf{n})] = \eta(\mathbf{p}) + tr, \quad \beta = (\beta_1 \quad \beta_2 \quad \dots \quad \beta_m)^T,$$

$\mathcal{A}$ is $q \times q$ matrix with elements

$$\mathcal{A}_{ij} = \begin{cases} \alpha_{ij}/2 \text{ for } i \neq j \\ \alpha_{ii} \text{ for } i = j \end{cases},$$

$\mathcal{B}$ is $m \times m$ matrix with elements

$$\mathcal{B}_{ij} = \begin{cases} \beta_{ij}/2 \text{ for } i \neq j \\ \beta_{ii} \text{ for } i = j \end{cases},$$

and $\mathcal{G}$ is $m \times q$ matrix with elements γ_{ij}, $i = 1,2,\dots,m;\ j = 1,2,\dots,q$.

As is shown by Vuchkov and Boyadjieva (1992) the following mean and variance models of the performance characteristic in mass production and usage can be induced on the basis of these models:

- Model of the mean value

$$\tilde{y} = E[y(\mathbf{p}+\mathbf{e},\mathbf{n})] = \eta(\mathbf{p}) + tr(\mathcal{B}\Sigma_{\mathbf{e}}) + tr(\mathcal{A}\Sigma_{\mathbf{n}}) = \\ = \beta_0 + \beta^T\mathbf{p} + \mathbf{p}^T\mathcal{B}\mathbf{p} + tr(\mathcal{B}\Sigma_{\mathbf{e}}) + tr(\mathcal{A}\Sigma_{\mathbf{n}}) \qquad (3)$$

- Variance model

$$\tilde{\sigma}^2 = \sigma^2[y(\mathbf{p}+\mathbf{e},\mathbf{n})] = (\beta + 2\mathcal{B}\mathbf{p})^T\Sigma_{\mathbf{e}}(\beta + 2\mathcal{B}\mathbf{p}) + \\ + (\alpha + \mathcal{G}^T\mathbf{p})^T\Sigma_{\mathbf{n}}(\alpha + \mathcal{G}^T\mathbf{p}) + HOT + \sigma_\varepsilon^2 \qquad (4)$$

where *HOT* denotes high order terms and can be computed by the formula

$$HOT = 2\sum_{i=1}^{m}\beta_{ii}^2\sigma_i^4 + \sum_{i=1}^{m-1}\sum_{j=i+1}^{m}\beta_{ij}^2\sigma_i^2\sigma_j^2 + 2\sum_{i=1}^{q}\alpha_{ii}^2\sigma_{ni}^4 + \\ + \sum_{i=1}^{q-1}\sum_{j=1}^{q}\alpha_{ij}^2\sigma_{ni}^2\sigma_{nj}^2 + \sum_{i=1}^{m}\sum_{j=1}^{q}\gamma_{ij}^2\sigma_i^2\sigma_{nj}^2. \qquad (5)$$

The regression model coefficients can be estimated through an experiment without errors in factor levels and substituted in (3) and (4). This way, one obtains estimates of the mean value and variance in mass production and usage of the product. These models can be used for quality improvement of the product or the process. Several definitions of the quality improvement problem can be given, one of them being: Minimize the variance while keeping the mean value on a target, τ.

Suppose that an experiment is carried out without errors in the factors. In this case one can use a general linear model of the form:

$$y(\mathbf{p},\mathbf{n}) = \mathbf{f}^T\theta + \varepsilon, \tag{6}$$

where **f** is nonrandom vector of the factors and θ is a vector of regression coefficients. For example model (2) can be written in the form (6) if **f** and θ are defined as follows:

$$\mathbf{f} = (1 \quad p_1 \quad \dots \quad p_m \quad p_1p_2 \quad \dots \quad p_{m-1}p_m \quad p_1^2 \quad \dots \quad p_m^2 \quad n_1 \quad \dots \quad n_q \\ n_1n_2 \quad \dots \quad n_{q-1}n_q \quad n_1^2 \quad \dots \quad n_q^2 \quad p_1n_1 \quad \dots \quad p_mn_q)^T$$

and

$$\theta = (\beta_0 \quad \beta_1 \quad \dots \quad \beta_m \quad \beta_{12} \quad \dots \quad \beta_{m-1,m} \quad \beta_{11} \quad \dots \quad \beta_{mm} \quad \alpha_1 \quad \dots \quad \alpha_q \\ \alpha_{12} \quad \dots \quad \alpha_{q-1,q} \quad \alpha_{11} \quad \dots \quad \alpha_{qq} \quad \gamma_{11} \quad \dots \quad \gamma_{mq})^T.$$

Though in the the product parameters **p** and the noise factors **n** can be set on given levels without errors in the experiment, in mass production **p** contains errors **e** and the elements of **n** are random. That is why the performance characteristic's value in mass production is

$$y(\mathbf{p}+\mathbf{e},\mathbf{n}) = \mathbf{h}^T\theta + \varepsilon, \tag{7}$$

where **h** is random vector, obtained from **f** through substitution of $p_i + e_i$ for p_i.

Denote $\mathbf{g} = \mathbf{h} - \mathbf{f}$ and $\psi = \mathbf{g} - E(\mathbf{g})$. Equation (7) can be written as follows:

$$y(\mathbf{p}+\mathbf{e},\mathbf{n}) = \mathbf{f}^T\theta + \mathbf{g}^T\theta + \varepsilon. \tag{8}$$

Taking into account that $\eta(\mathbf{p}) = \mathbf{f}^T\theta$ is nonrandom, one can compute the mean value of the performance characteristic in mass production as follows:

$$\tilde{y} = E[y(\mathbf{p}+\mathbf{e},\mathbf{n})] = \eta(\mathbf{p}) + \theta^T E(\mathbf{g}). \tag{9}$$

Denote also $\psi = \mathbf{g} - E(\mathbf{g})$ and $\Psi = E(\psi\psi^T)$. As $\mathbf{f}^T\theta$ is not random one can compute the variance as follows:

$$\tilde{\sigma}^2 = \sigma^2[y(\mathbf{p})] = \sigma^2(\mathbf{g}^T\theta) + 2\operatorname{cov}(\mathbf{g}^T\theta,\varepsilon) + \sigma_\varepsilon^2. \tag{10}$$

Usually $\mathbf{e}$ and ε are uncorrelated and $2\operatorname{cov}(\mathbf{g}^T\theta,\varepsilon) = 0$. Consequently the variance of the performance characteristic in mass production is

$$\tilde{\sigma}^2 = E(\theta^T\psi\psi^T\theta) + \sigma_\varepsilon^2 = \theta^T\Psi\theta + \sigma_\varepsilon^2. \tag{11}$$

Models (3) and (4) are special cases of (9) and (11).

Some authors use models in which they do not make difference between errors in the product parameters and external noise factors. We consider this difference as important one by two reasons:

- If the moments of the errors in product parameters are known one does not need to vary these noise factors during the experiments (Vuchkov and Boyadjieva (1989)). This simplifies the experimental design and reduces the number of experiments.
- The properties of the models with errors in product parameters and these with both errors in the parameters and external noise factors are different (Vuchkov and Boyadjieva (1988), Vuchkov and Boyadjieva (1992)). That is why the optimization procedures in these two cases are different.

The mean and variance models provide correct predictions if the assumptions used in their derivation are met. In many practical applications there are some deviations that can cause inaccurate predictions. In this paper we consider three main sources of model inaccuracy as follows:

- Incorrect regression model structure,
- Neglect of the high order distribution moments and the high order terms in the mean and variance models,
- Inaccuracy of the regression model estimation.

2. Inacccuracy due to incorrect regression model structure

Suppose that the true model structure is $\eta = \mathbf{f}^T\theta$, while the postulated model is less terms than the true one and is $\eta_1 = \mathbf{f}_1{}^T\theta_1$, where $\mathbf{f}^T = (\mathbf{f}_1^T \vdots \mathbf{f}_2^T)$ and $\theta^T = (\theta_1^T \vdots \theta_2^T)$. The vectors $\mathbf{g}$ and $\psi = \mathbf{g} - E(\mathbf{g})$ can be presented in block form as follows:

$$\mathbf{g} = (\mathbf{g}_1^T \vdots \mathbf{g}_2^T) \text{ and } \psi^T = (\psi_1^T \vdots \psi_2^T).$$

The matrix Ψ can be written in the following block form:

$$\Psi = E(\psi\psi^T) = \begin{bmatrix} \Psi_{11} & \Psi_{12} \\ \Psi_{21} & \Psi_{22} \end{bmatrix},$$

where $\Psi_{ij} = \psi_i\psi_j^T; \quad i, j = 1,2.$

Assuming that e_i, n_i and ε are independent, the variance can be computed by (11) as follows:

$$\tilde{\sigma}^2 = \theta^T\Psi\theta + \sigma_\varepsilon^2 = \theta_1{}^T\Psi_{11}\theta_1 + \theta_2{}^T\Psi_{22}\theta_2 + 2\theta_1{}^T\Psi_{12}\theta_2 + \sigma_\varepsilon^2 \,. \quad (12)$$

The variance obtained through the postulated model is following:

$$\tilde{\sigma}_1^2 = \theta_1{}^T\Psi_{11}\theta_1 + \sigma_\varepsilon^2.$$

Hence $\tilde{\sigma}_1^2$ is biased and the bias is equal to

$$\Delta\sigma^2 = \theta_2{}^T\Psi_{22}\theta_2 + 2\theta_1{}^T\Psi_{12}\theta_2\,. \quad (13)$$

Therefore if some of the significant factors or functions f_i are not in the model their error causes a bias $\Delta\sigma^2$ in the variance estimate. The model (12) can be written in the form

$$\tilde{\sigma}^2 = \theta_1{}^T\Psi_{11}\theta_1 + \sigma_N^2\,, \quad (14)$$

where $\sigma_N^2 = \Delta\sigma^2 + \sigma_\varepsilon^2$ is variance of some noise incorporating all noise components which are not taken into account in the model of the variance $(\sigma_N^2 \geq \sigma_\varepsilon^2)$.

The variance σ_N^2 depends on the regression coefficients and on the functions **f** which are included in the term $2\theta_1{}^T\Psi_{12}\theta_2$ through Ψ_{12}. Therefore if the postulated model has less terms than necessary then σ_N^2 can not be considered as a constant. In any case the observations are heteroscedastic but the true model of variance explains the sources of heteroscedasticity, while a model with incorrect structure incorporates them into a single term σ_N^2, which can not be used for making decisions.

An important question which should be addressed is how good are the models of mean value and variance in mass production. Starting this discussion we should not forget that these models are derived *analytically* on the basis of a regression model. The experiments used for estimation of the regression are without errors in product parameters, while models (3) and (4) provide predicted mean values and variances in mass production where the errors can not be eliminated. Therefore the values predicted by these models should not be compared with the experimental data used for derivation of the regression equation. The observed optimal parameter values can be verified through a confirmatory experiment. Several observations should be made for the optimal parameter values and the obtained sample mean and variance should be compared with the values predicted by models (3) and (4).

The quality of mean value and variance models in mass production depends on goodness-of-fit for the regression model. For a high quality regression model and good estimates of noise variances the models (3) and (4) provide good prediction of the mean value and variance in mass production if the assumptions for their derivation are met. That is why the attention of the engineer or the scientist should be focused on obtaining a high quality regression model using the recommendations for design of experiments and statistical data analysis. The assumptions about the noise distributions should also be checked. If there is no evidence that the noise is normally distributed formula (15) of the next section should be used for computation of the performance characteristic's variance.

3. Inaccuracy due to neglecting the high order error distribution moments and the high order terms in the variance model

The mean value of the performance characteristic in mass production is obtained by taking expectation of the response predicted by a regression model of order l. Equation (9) shows that the highest order of error distribution moments in $\tilde{y}$ is also l because the errors are included in vector ψ through the same functions **f** which are used in the regression. For example, one can see from (3) that for a second order polynomial model (1) the mean value $\tilde{y}$ contains second order error distribution moments $\left(\sigma_i^2\right)$. If the error distribution is symmetrical (for example normal) then the third order moments are equal to zero. That is why for symmetrical distributions the model of mean value in mass production and usage obtained on the basis of third order polynomial regression depends only on the error variances σ_i^2. Hence $\tilde{y}$ does not depend on high order error distribution moments for second and third order polynomial regression models.

By definition the variance is a quadratic transformation of a random variable. Therefore if the degree of a polynomial regression model is l then the highest order of error distribution moments included in the model for variance in mass production is $2l$. Therefore if l = 2 and the error distributions are symmetrical then the model (4) takes into account all necessary moments. For normal distribution all high order moments can be computed through variances and consequently in this case only variance estimates are needed.

If the error distributions are not symmetrical then the formula for the variance is changed. For example for the case with errors only in product parameters the variance model is

$$\tilde{\sigma}^2 = \sigma^2\left[y(\mathbf{p})\right] = \sum_{i=1}^{m} \sigma_i^2 \left(\beta_i + 2\beta_{ii} p_i + \sum_{j=1,\, j \neq i}^{m} \beta_{ij} p_j \right)^2 + \sum_{i=1}^{m} \beta_{ii}^2 \left(\mu_{i4} - \sigma_i^4 \right) +$$

$$\hat{\theta} = \left(\hat{\theta}_1 \;\; \hat{\theta}_2 \;\; \ldots \;\; \hat{\theta}_k\right)^T \; \theta = \left(\theta_1 \;\; \theta_2 \;\; \ldots \;\; \theta_k\right)^T \qquad (15)$$

In this case third and fourth distribution moments (μ_3 and μ_4) are needed for variance prediction. It can also be shown that if the regression model is of third order then the high order terms in variance equation contain the product parameters. Hence the neglect of high order distribution moments may result in non optimal parameter values.

We can not recommend neglecting the high order error distribution moments in a model of variance without a careful study of their influence. Consider for example in more detail the case of second order regression model. The high order terms in the model of variance in mass production may not be negligible in the following cases:

•*The error distribution is not normal or is at least non symmetrical.* One can see from (15) that in this case the high order terms of the variance model depend on third and fourth error distribution moments μ_{i3} and μ_{i4}. In this equation μ_{i3} is multiplier of an expression containing product parameters. Hence the optimal parameter values depend on third order moments.

•*The error variance depends on the values of the product parameters.* This is typical for the case when the tolerance intervals or component parts depend on the nominal parameter values. Other reasons for such dependence can also exist.

4. Inaccuracy due to random character of the regression coefficients estimates

In Section 1 we assumed that the regression model coefficients in (1), (2) and (6) are known constants. In fact only estimates $\hat{\theta} = \left(\hat{\theta}_1 \quad \hat{\theta}_2 \quad \ldots \quad \hat{\theta}_k\right)^T$ are available and they are substituted in the models for $\theta = \left(\theta_1 \quad \theta_2 \quad \ldots \quad \theta_k\right)^T$. This causes some variability of the predicted mean value $\tilde{y}$ and variance $\tilde{\sigma}^2$. Consider equation (6). Suppose that the factor levels are set without errors in the experiment. The least squares estimates of the regression coefficients are

$$\hat{\theta} = \left(\mathbf{F}^T\mathbf{F}\right)^{-1}\mathbf{F}^T\mathbf{y},$$

where $\mathbf{F}$ is $N \times k$ matrix of regressors. Putting $\hat{\theta}$ for θ in the model of the mean value in mass production (5.36) one obtains

$$\hat{\tilde{y}} = \hat{y}(\mathbf{p}) + \boldsymbol{\theta}^T E(\mathbf{g}).$$

It is not difficult to prove that $\hat{\tilde{y}}$ is an unbiased estimate of $\tilde{y}$ because $E(\mathbf{g})$ does not depend on the output noise ε, while $\hat{y}(\mathbf{p})$ and $\hat{\theta}$ are unbiased estimates of $\eta(\mathbf{p})$ and θ correspondingly. Actually taking expectation with respect to ε one obtains:

$$E_\varepsilon\left[\hat{\tilde{y}}\right] = E_\varepsilon\left[\hat{y}(\mathbf{p})\right] + E_\varepsilon\left[\hat{\theta}^T E(\mathbf{g})\right] = \eta(\mathbf{p}) + \theta^T E(\mathbf{g}) = \tilde{y}.$$

Under the assumptions that ε and e_i are uncorrelated an estimate $\tilde{s}^2$ of $\tilde{\sigma}^2$ is obtained substituting $\hat{\theta}$ foθ and s_ε^2 for σ_ε^2 in (11) as follows:

$$\tilde{s}^2 = \hat{\theta}^T \Psi \hat{\theta} + s_\varepsilon^2. \tag{16}$$

The first term in the right hand side of (16) can be presented in the form

$$\hat{\theta}^T \Psi \hat{\theta} = \left(\hat{\theta} - \theta\right)^T \Psi \left(\hat{\theta} - \theta\right) + 2\hat{\theta}^T \Psi \theta - \theta^T \Psi \theta. \tag{17}$$

The expectation of $\tilde{s}^2$ with respect to the output noise ε is computed putting (17) into (16) and taking into account that $\hat{\theta}$ are unbiased estimates: $E_\varepsilon\left(\hat{\theta}\right) = \theta$. Following relationship is obtained:

$$E_\varepsilon\left(\tilde{s}^2\right) = \theta^T \Psi \theta + \sigma_\varepsilon^2 + E_\varepsilon\left[\left(\hat{\theta} - \theta\right)^T \Psi \left(\hat{\theta} - \theta\right)\right] = \tilde{\sigma}^2 + tr\left[\Psi \mathbf{V}\left(\hat{\theta}\right)\right], \tag{18}$$

where $\mathbf{V}\left(\hat{\theta}\right) = E_\varepsilon\left[\left(\hat{\theta} - \theta\right)\left(\hat{\theta} - \theta\right)^T\right]$ is $k \times k$ covariance matrix with elements

$$v_{ij} = \begin{cases} \sigma^2\left(\hat{\theta}_i\right) \text{ for } i = j \\ \operatorname{cov}\left(\hat{\theta}_i, \hat{\theta}_j\right) \text{ for } i \neq j \end{cases}.$$

Equation (18) shows that $\tilde{s}^2$ is a biased estimate of $\tilde{\sigma}^2$. As Ψ is a non negative definite matrix it follows from (18) that $E_\varepsilon\left(\tilde{s}^2\right) \geq \tilde{\sigma}^2$. That means that the substitution of $\hat{\theta}$ foθ causes an increase of $\tilde{s}^2$ as compared with $\tilde{\sigma}^2$.

Taking into account that Ψ is independent of ε this bias can be estimated. Equation (18) can be written in the form

$$E_\varepsilon(\tilde{s}^2) = \tilde{\sigma}^2 + \sum_{l=1}^{k} \Psi_{ii}\sigma^2(\hat{\theta}_i) + 2\sum_{l=1}^{k-1}\sum_{t=l+1}^{k} \Psi_{lt} \operatorname{cov}(\hat{\theta}_l, \hat{\theta}_t), \qquad (19)$$

where Ψ_{ll} and Ψ_{lt} are elements of Ψ.

As the estimates $\hat{\theta}$ are computed by the least squares method, their covariance matrix is defined by

$$\mathbf{V}(\hat{\theta}) = (\mathbf{F}^T\mathbf{F})^{-1}\sigma_\varepsilon^2 = \mathbf{C}\sigma_\varepsilon^2,$$

$$\text{or } \sigma^2(\hat{\theta}_l) = c_{ll}\sigma_\varepsilon^2 \text{ and } \operatorname{cov}(\hat{\theta}_l, \hat{\theta}_t) = c_{lt}\sigma_\varepsilon^2,$$

where c_{ll} and c_{lt} are diagonal and off-diagonal elements of **C** correspondingly. Putting these equations in (19) one obtains

$$E_\varepsilon(\tilde{s}^2) = \tilde{\sigma}^2 + \sum_{l=1}^{k} \Psi_{ll}c_{ll}\sigma_\varepsilon^2 + 2\sum_{l=1}^{k-1}\sum_{t=l+1}^{k} \Psi_{lt}c_{lt}\sigma_\varepsilon^2 .$$

The bias is often negligible. Substituting s_ε^2 for σ_ε^2 and $\tilde{s}^2$ for $\tilde{\sigma}^2$ following *unbiased* estimate of the variance of the performance characteristic in mass production is obtained:

$$\hat{s}^2 = \hat{\theta}^T\Psi\hat{\theta} + s_\varepsilon^2\left(1 - \sum_{l=1}^{k} \Psi_{ll}c_{ll} - 2\sum_{l=1}^{k-1}\sum_{t=l+1}^{k} \Psi_{lt}c_{lt}\right). \qquad (20)$$

When the number of observations increases, the elements of **C** often decrease. For large number of observations and small value of s_ε^2 the bias can be neglected.

References

Barker, T. and Clausing, D. (1984). Quality engineering by design. The Taguchi method. *The international QC forum*, **11**, No. 8, 28-52.

Box, G. E. P. and Fung, C. A. (1986). *Studies in quality improvement: mimimizing transmitted variation by parameter design.* Report No. 8, University of Wisconsin - Madison, Center for Quality and Productivity Improvement.

Box, G. E. P. and Jones, S. (1992a). Designing products that are robust to environment. *Total Quality Management*, **3**, No. 3, 265 - 282.

Vining, G. G. and Myers, R. H. (1990). Combining Taguchi and response surface philosophies: a dual response approach. *Journal of Quality Technology*, **22**, 38 - 45.

Vuchkov, I. N., L. N. Boyadjieva (1981). Errors in the factor levels and estimation of regression model parameters. *J. Statist. Comput. Simul.* **13**, 1-12.

Vuchkov, I. N., L. N. Boyadjieva (1988). The robustness against tolerances of performance characteristics described by second order polynomials. In *Optimal Design and Analysis of Experiments.* Eds. Y. Dodge, V. Fedorov, H. P. Wynn, Elsevier, Amsterdam, 293 - 309.

Vuchkov, I. N. and Boyadjieva, L. N. (1989). A model-based approach to the robustness against tolerances. *Proceedings of 33rd EOQC Annual Conference*, Vienna, 585-592.

Vuchkov , I. N. and Boyadjieva, L. N. (1990a). Model based quality improvement. *Proceedings of the seminar on systematical statistical analysis of processes.* Varna, 110-125.

Vuchkov, I. N. and Boyadjieva, L. N. (1990b). Taguchi method and model based quality improvement - a comparison. *Proceedings of 34rd EOQC Annual Conference*, Dublin, 383 - 390.

Vuchkov, I. N., L. N. Boyadjieva (1992). Quality improvement through design of experiments with both product parameters and external noise factors. In *Model Oriented Data Analysis.* Eds. V. Fedorov, W. G. Müller, I. N. Vuchkov. Physica-Verlag, Heidelberg, 195-212.

Vuchkov, I. N., L. N. Boyadjieva (1994). Computer-aided quality improvement. *Proceedings 3-rd IEEE Conference on Control Applications*, **2**, Glasgow, 1265-1269.

Vuchkov, I. N., L. N. Boyadjieva, P. S. Damianov (1987). On the design of experiments for quality improvement. *Satellite conferenc of the 17th European Meeting of Statisticians*, Thessaloniki, p. 23.

Part II

Reliability and Failure Time Analysis

Accelerated Life Test with Some Parameter Change in Life-Stress Relation

WEN-TAO HUANG
Institute of Statistic Science
Academia Sinica
Taipei, Taiwan, R.O.C.

HUEI-TSAN LIN
Institute of Management Science
Tamkang University
Taipei, Taiwan, R.O.C.

Key Words - Accelerated life test, censored data, change point.

Abstract - In most literature, accelerated life testing (ALT) models always assume the parameters in the life-stress relation remain unchanged under all stress levels. But in many practical situations, the upper bound of the stress levels under which such parameters keep unchanged is often unknown. In this paper we study the situation that the life-stress relation keeps unchanged, however, some parameters involved may be changed if it is over-stressed. A general method is propsed to consider such situations.

1. INTRODUCTION

Accelerated life tests (ALT) are nowadays widely used for experiments on many industrial products. ALT can be used to reduce the failure time on tests to save a lot of time and money, and often the data are censored because of high reliability products. Theory and applications for various types of ALT have been studied in detail in the widely circulated book by Nelson [16].

Usually, it is assumed that the domain of stress levels is known so that the life-stress relation keeps unchanged as long as the level is in this domain. For a new product, or a new design of several components, the situations may not be so. In this paper the linearity of the transformed life-stress relation is assumed to hold, but its parameters involved in the linear relation may change when the stress exceeds some unknown bound.

Consider the situation that the life-stress relations can be transformed into linear relations. The most two famous examples of such situation are the Arrhenius and inverse power law models (for details, see Nelson [16]). They belong to log-linear relationship. If we consider hazard rate instead of life distribution and consider some associated covariate as stress, then the well-known Cox model [4] belongs to this class. Liang et al [12] studied the effect of reactivity to the cold pressor test and examined the hypothesis that the effect may vary with early versus late onset of hypertension through a set of data gathered on a group of white male medical students. It is found that the Cox proportional hazards model keeps unchanged,

*Kitsos, C.P. and Edler, L., Eds.:Industrial Statistics, Physica-Verlag, Heidelberg, 1997

however, there exists a change point at 40 for the covariate of age. So considering age as a kind of stress to life, in the model for a prospective epidemio epidemiological study of hypertension, the relationship of hazard rate and hypertension keeps log-linear, however, there is a two-phase regression with respect to covariate of age at 40. Other example may also refer to the life distribution for patients with leukemia after remission induction. In this study, Matthews and Farewell [13] consider exponential life distribution and find out that there is a change in scale parameter after induction of remission.

If life-stress relation changes other than parameters change when stress level exceeds some bound, for example, from log-linear to log-nonlinear, the problem is rather difficult and it may involve model selection problem and which is out of scope of this paper. So far, there is still very few justification for this situation.

The Arrhenius model is widely used for product life as a function of temperature. Applications include electrical insulations and dielectrics, solid state and semiconductor devices, battery cells, lubricants and greases, plastics and incandescent lamp filaments. For inverse power law model, it is applied in electrical insulations in voltage-endurance tests, ball and roller bearings, metal fatigue due to mechanical loading and thermal cycling etc.. These two models can be expressed respectively by

$$\theta = A \cdot \exp[B/l] \tag{1}$$

$$\theta = C/l^D \tag{2}$$

where θ is some parameter in life distribution and l is stress level under which the test is carried out and $A > 0$, B, $C > 0$ and D are some constants. Taking logarithm on both sides of these equations, we have a linear life-stress relation

$$\mu = \beta_0 + \beta_1 x \tag{3}$$

where $x = 1/l$ for the Arrhenius model and $x = -\ln l$ for the inverse law and μ is some function of θ, β_0 and β_1 are constants depending on A and B or C and D.

In usual models, it is assumed that (3) holds for either $x > x_0$ or $x \leq x_0$ for some known x_0. Generally, it may not be true that (3) holds for all $x > 0$.

In this paper, we consider models such that the linear life-stress relation (3) holds through some transformation for which the unknown parameters β_0 and β_1 are permitted to have change when $x > x_0$ for some unknown

x_0. To be specific, we consider the following model

$$\mu = \begin{cases} \beta_0 + \beta_1 X & x \leq x_0 \\ \beta_0' + \beta_1' X & x > x_0 \end{cases} \tag{4}$$

where x_0 is some unknown value.

We propose a methed to test if there is some change of β_0 and β_1. In both cases, we propose method to estimate the unknown parameters of the life-stress relation. We specially focus on the Weibull and Lognormal life models (see Lawless [11], Crow and Shimizu [5]) based on Type I or Type II censored data. For other situations, the proposed methods are analogously applicable.

Section 2 describes our model assumptions and a related two-phase linear regression. Section 3 illustrates examples for Weibull and Lognormal lifetime distributions. Some simulation results are also given.

2. MODEL ASSUMPTIONS

(1) Each time of experiment, a constant stress is applied to test and the functional form of life-stress relation keeps unchanged, however, unknown parameters involved in the relation is permitted to change at most once.

(2) There are m stress levels, say, $l_1 < \cdots < l_m$. Under each stress level l_i, the life distribution of item for test is the same, except the parameters involved may change. There exists some transformation that the life-stress relation through this transformation may be represented by a linear relationship given by

$$\mu_i = \beta_0 + \beta\ x_i, \qquad i = 1, \cdots, m, \tag{5}$$

where x_i is some known function of l_i and μ_i may involve l_i and parameters in the life distribution of item for test, however μ_i can be estimated from data, β_0 and β_1 are some unknown parameters.

(3) When the stress level l_i exceeds some unknown level, say l_{i_0}, the parameters β_0 and β_1 may change, however at most one change is permitted and the model (4) is assumed.

(4) Data can be collected under various censoring scheme.

When the life distribution is Weibull with scale parameter θ_i and shape parameter δ_i under stress level l_i, the Arrhenius-Weibull model can be represented by (5) where

$$\mu_i = \ln \theta_i, \; x_i = 1/l_i \tag{6}$$

and keeping δ_i constant, $i = 1, \cdots, m$. When it is Weibull-Power model, everything is the same except $x_i = -\ln l_i$. Situations for Lognormal remain the same with mean $\ln \theta_i$ and constant variance δ_i.

For its conveinience of illustration , we focus on two life distributions, Weibull and Lognormal respectively with inverse power law models.

If μ_i is estimated, denote it by y_i, the model of (4) can be rewritten by the following matrix form,

$$\underset{\sim}{y} = \begin{bmatrix} X_1 & 0 \\ 0 & X_2 \end{bmatrix} \cdot \begin{bmatrix} \underset{\sim}{\beta} \\ \underset{\sim}{\beta}' \end{bmatrix} + \underset{\sim}{\varepsilon}. \tag{7}$$

where $x_k \leq x_0 < x_{k+1}$ and x_k and x_{k+1} are respectively involved in X_1 and $X_2, \underset{\sim}{\varepsilon}$ are error terms due to estimates. When MLE are used to estimate $\mu_i, \underset{\sim}{\varepsilon}$ are multivariate normal with mean $\underset{\sim}{0}$ and covariance matrix $diag(\sigma_1^2, \cdots, \sigma_m^2)$ which can be estimated.

If there is no change, least squares methods can be used to estimate β_0 and β_1 in model (5) and there are many papers devoted this problem (see, for example, Schmee and Hahn [17]).

If there exists a change, i.e. if the model (7) holds, it needs to estimate k so that least squares method can be applied again. Therefore, under model (7), it needs, first of all, to test $H_0 : \underset{\sim}{\beta} = \underset{\sim}{\beta}'$ versus $H_1 : \underset{\sim}{\beta} \neq \underset{\sim}{\beta}_1'$.

If H_0 is accepted, we apply the usual least square estimators for $\underset{\sim}{\beta}$. If H_0 is rejected, we need to estimate k, so that again we can apply the usual least-square method to estimate $\underset{\sim}{\beta}$ and $\underset{\sim}{\beta}'$. And thus θ_0, the parameter under design stress can be estimated through the relation (5).

It is clear that equation (7) is a two-phase linear regression model. There are many papers devoted to the study of such two-phase linear regression testing problems. Among others, Worsley [18] provides the upper bounds for null exceedence probabilities for the likelihood ratio statistic. Hinkley ([8],[9]) concerns this testing problem with continuous constraint at the change point. Jandhyala and Macneill [10] provides an unified Bayesian approach to the problem of testing for parameter changes at unknown times in linear regression models, however, this Bayesian approach always assumes some additional distributions on μ_i and change point x_0 which are not always available. Note that all the methods previously mentioned need to assume $\underset{\sim}{\varepsilon} \sim N(\underset{\sim}{0}, \sigma^2 I_m)$. If the distribution of $\underset{\sim}{\varepsilon}$ is unknown, then some nonparametric procedures can be used, which can be referred to Deshayes, Picard [6] and Fomby et al.[7].

3. EXAMPLES

To illustrate our model, we consider models of the Weibull and Lognormal distributions with Type I or Type II censored data. All test units start and run simultaneously. For Type I censoring, the test continues until a fixed length of time. And for Type II censoring, only the first r_i failure times of the n_i tested products are observed under l_i, $i = 1, \cdots, m$. We consider these two models in separate parts.

3.1A WEIBULL LOG-LINEAR MODEL

Let T_{ij} denote the life time of the jth item under stress level $l_i, i = 1, \cdots, m, j = 1, 2, \cdots, n_i$. Assume that T_{ij} follows the Weibull distribution with scale parameter θ_i and shape parameter δ. Then lnT_{ij} follows the extreme value distribution with location parameter $\mu_i = ln\theta_i$ and scale parameter $\eta = 1/\delta$.

Under Type I censoring, maximum likelihood estimates (MLE) for μ_i and η can be obtained by solving simultaneous equations $\dfrac{\partial L}{\partial \mu_i} = 0$ and $\dfrac{\partial L}{\partial \eta} = 0$, where L is the likelihood function, and numerically, the Newton-Raphson method can be applied (see Lawless [11] and Bugaighis [1]). It has been shown these MLE for μ_i and η are asymptotically bivariate normal with mean μ_i and η and covariance I_0^{-1}, where I_0 is the observed information matrix. Meeker and Nelson [14] have tabulated the asymptotic covariance matrices of the MLE.

Under Type II censoring, suppose test continues until the r_ith failure occurs. MLE for μ_i and η have been studied by Cohen [2,3]. Estimates of variances of the MLE can also be computed. Asymptotic normality of the MLE assures the model (7) and the variances of $\underset{\sim}{\varepsilon}$ can be obtained.

Finally, it is to be noted, when the inverse-power law is assumed, $x_i = -lnl_i$ in (4) or in (7).

3.1B LOGNORMAL LOG-LINEAR MODEL

Let T_{ij} denote the life time of the jth item under stress $l_i, i = 1, \cdots, m, j = 1, 2, \cdots, n_i$. If $logT_{ij}$ follows the normal distribution with mean $ln\theta_i$, variance δ_i^2, then T_{ij} follows the Lognormal distribution with corresponding parameters. The life-stress relation is also given by (3) with (6) assuming that $\delta_i^2 = \delta^2$ is independent of stress.

For Type I censoring sampling, the Newton-Raphson method can be applied to solve for MLE for the parameters $\mu_i = ln\theta_i$ and δ^2 (see Lawless [12]).

For Type II censor sampling, Cohen [2] considered the MLE for μ_i and δ^2.

It is noted that the life-stress relationship defined by (3) with (6) are also valid for either Arrhenius or inverse power law.

3.2 TESTING

For given data T_{ij} either under Type I or Type II censoring, model (7) can be set up through the discussion given in 3.1A and 3.1B where $\underset{\sim}{y}$ are all estimates of μ_i and $\underset{\sim}{\varepsilon}^t = (\varepsilon_1, \varepsilon_2, \cdots, \varepsilon_m)$ are independently normally distributed with all mean 0 and variance σ_i^2 respectively. These σ_i^2 can be computed or estimated as mentioned in 3.1. Let $\underset{\sim}{P}$ denote the computed or estimated vector of inverse of standard deviations, i.e. $\underset{\sim}{P} = diag(1/\sigma_1, \cdots, 1/\sigma_m)$. Multiply both side of (7) by $\underset{\sim}{P}^t$. Then, (7) becomes

$$\begin{bmatrix} \underset{\sim}{y}_1^* \\ \underset{\sim}{y}_2^* \end{bmatrix} = \begin{bmatrix} \underset{\sim}{X}_1^* & 0 \\ 0 & \underset{\sim}{X}_2^* \end{bmatrix} \cdot \begin{bmatrix} \underset{\sim}{\beta} \\ \underset{\sim}{\beta}' \end{bmatrix} + \begin{bmatrix} \underset{\sim}{\varepsilon}_1^* \\ \underset{\sim}{\varepsilon}_2^* \end{bmatrix}. \tag{8}$$

where

$\underset{\sim}{y_1}^* = (y_1^*, \cdots, y_k^*)^t, \underset{\sim}{y_2}^* = (y_{k+1}^*, \cdots, y_m^*)^t, \underset{\sim}{\beta} = (\beta_0, \beta_1)^t, \underset{\sim}{\beta}' = (\beta_0', \beta_1')^t,$

$$\underset{\sim}{X_1}^* = \begin{bmatrix} 1/\sigma_1, & \cdots, & 1/\sigma_k \\ x_1/\sigma_1, & \cdots, & x_k/\sigma_k \end{bmatrix}, \quad \underset{\sim}{X_2}^* = \begin{bmatrix} 1/\sigma_{k+1}, & \cdots, & 1/\sigma_m \\ x_{k+1}/\sigma_{k+1}, & \cdots, & x_m/\sigma_m \end{bmatrix}$$

$\underset{\sim}{\varepsilon_1}^* = (\varepsilon_1/\sigma_1, \cdots, \varepsilon_k/\sigma_k)^t, \quad \underset{\sim}{\varepsilon_2}^* = (\varepsilon_{k+1}/\sigma_{k+1}, \cdots, \varepsilon_m/\sigma_m)^t.$

It is obvious that both $\underset{\sim}{\varepsilon_1}^*$ and $\underset{\sim}{\varepsilon_2}^*$ have zero mean vector $\underset{\sim}{0}$ and identity matrices. If k is given, the MLE of $\underset{\sim}{\beta}$ and $\underset{\sim}{\beta}'$, denoted by $\hat{\underset{\sim}{\beta}}(k)$ and $\hat{\underset{\sim}{\beta}}'(k)$ are respectively given by

$$\hat{\underset{\sim}{\beta}}(k) = (\underset{\sim}{X_1}^{*t}\underset{\sim}{X_1}^*)^{-1}\underset{\sim}{X_1}^{*t}\underset{\sim}{y_1}^*, \quad \hat{\underset{\sim}{\beta}}'(k) = (\underset{\sim}{X_2}^{*t}\underset{\sim}{X_2}^*)^{-1}\underset{\sim}{X_2}^{*t}\underset{\sim}{y_2}^* \tag{9}$$

Take $\underset{\sim}{\Delta} = \underset{\sim}{\beta}' - \underset{\sim}{\beta}$. Then, to test $H_0 : \underset{\sim}{\beta} = \underset{\sim}{\beta}'$ is equivalent to test $H_0' : \underset{\sim}{\Delta} = \underset{\sim}{0}$. For given k, the MLE for $\underset{\sim}{\Delta}$ is given by $\hat{\underset{\sim}{\Delta}}(k) = \hat{\underset{\sim}{\beta}}'(k) - \hat{\underset{\sim}{\beta}}(k)$ and its variance is given $var(\hat{\underset{\sim}{\Delta}}(k)) = (\underset{\sim}{X_1}^{*t}\underset{\sim}{X}_1^*)^{-1} + (\underset{\sim}{X_2}^{*t}\underset{\sim}{X}_2^*)^{-1}$. Since $\underset{\sim}{y_1}^*(k)$ and $\underset{\sim}{y_2}^*(k)$ are independent for each k, thus $-2ln$ (likelihood ratio) for fixed k is given by $U_k = \hat{\underset{\sim}{\Delta}}^t(k) var(\hat{\underset{\sim}{\Delta}}(k)) \hat{\underset{\sim}{\Delta}}(k)$, which is the reduction in residual sum of squares under H_1 (not H_0). Let $U = \max_{2 \le k \le m-2} U_k$. Then, the approximate upper bound of U under H_0' (i.e. H_0) is given by the following inequality (see Worsely [18]).

$$\begin{aligned} P(U \ge u) \;&\le\; [1 - F_p(u) + (2/\pi) f_{p+1}(u)] \\ &\times [\sum_{k=p}^{m-p-1} \tau_k - \{u/(2p+2) - 1\} \cdot \sum_{k=p}^{m-p-1} \tau_k^3/6]. \end{aligned} \tag{10}$$

If some of $\sigma_1, \cdots, \sigma_m$ are unknown, or δ^2 is unknown and is estimated, then we can use $\tilde{\underset{\sim}{P}}$ to estimate $\underset{\sim}{P}$ using consistent estimator of σ_i, $i =$

$1, \cdots, m$. Under this situation, instead of U, we use $V = U/Q$, where Q is the residual sum of squares under H_0. Then, we have

$$P(V > v) \leq M(v) \approx 1 - G_{p,\lambda}(v) + (2/\pi) g_{p+1,\lambda+1}(v)$$
$$/(m-p)[\sum_{k=p}^{m-p-1} \tau_k - \{\frac{(\lambda-1)v}{(2p+2)(1-v)} - 1\} \sum_{k=p}^{m-p-1} \tau_k^3/6] \qquad (11)$$

Reject H_0 when $U(V)$ is large. Critical values of u or v for rejecting H_0 associated with various Type I errors can be computed from (10) or (11). However, the test is conservative (real α-value will be less), since the right hand side of (10) is an upper bound.

If H_0 is accepted, just apply the usual least squares estimate for $\underset{\sim}{\beta}$. However, if H_0 is rejected, find k so that $U_k = U$. Then $\underset{\sim}{\beta}$ and $\underset{\sim}{\beta}'$ are respectively estimated by $\hat{\underset{\sim}{\beta}}(k)$ and $\hat{\underset{\sim}{\beta}}'(k)$ given by (9).

3.3. ONE-STAGE PROCEDURE

Instead of considering a change-point model in linear regression given by (7), we can also directly estimate the unknown parameters by maximum likelihood principle which has been described in details in Nelson [16] (pp.284-301). Based on the sample (log)likelihood of a set of data and using the life-stress relation (5), we can estimate the unknown parameters β_0 and β_1 by maximizing the sample (log)likelihood. Various likelihood function based on Type I or Type II censored data can be found in Lawless [11] among others.

Let L denote the likelihood function, then it is a function of lifetime (observerd data) and unknown parameters μ_i and δ (for the case of Weibull and Lognormal). So, $L = \prod_{i=1}^{m} f(t_i; \mu_i, \delta)$ which can be rewritten by $L = \prod_{i,j} f(t_{ij}; \beta_0 + \beta_1 x_i, \delta)$ through life-stress relation. By maximum likelihood principle, the parameters β_0, β_1 and δ can be estimated and let them be denoted by $\hat{\beta}_0, \hat{\beta}_1$ and $\hat{\delta}$ respectively. Then, $(\hat{\beta}_0, \hat{\beta}_1, \hat{\delta})$ is asymptotically multivariate normal under some regularity conditions. The sample Fisher information matrix and thus the covariance matrix can be estimated (see, for example, Nelson[16] p.292). Therefore, point and interval estimates for β_0, β_1 and δ can be obtained.

When there is a parameter change, the situation is different and it can be represents by

$$\begin{aligned} \mu_i &= \beta_0 + \beta_1 x_i & i &= 1, 2, ..., k \\ &= \beta_0' + \beta_1' x_i & &= k+1, ..., m \end{aligned}$$

where k is unknown parameter. We need to estimate k so that the data set is separated into two parts where the first part consists of all these data corresponding up to level k and the second part consists of the remaining. Using the likelihood functions they are respectively given by $\prod_{i=1}^{k} f(t_i; \beta_0+\beta_1 x_i, \delta)$ and $\prod_{j=k+1}^{m} f(t_j; \beta_0'+\beta_1' x_j, \delta')$. Let the ML estimates be denoted by $\hat{\beta}_0(k), \hat{\beta}_1(k), \hat{\delta}(k)$ and $\hat{\beta}_0'(k), \hat{\beta}_1'(k), \hat{\delta}'(k)$ respectively. If there is a change at k and k is correctly estimated, then $(\hat{\beta}_0(k), \hat{\beta}_1(k), \hat{\delta}(k))$ is asymptotically multivariate normal and so is $(\hat{\beta}_0'(k), \hat{\beta}_1'(k), \hat{\delta}'(k))$. Unfortunately, there is no results can be applied to test for a change nor known method can be used to estimate k. This question in general is not simple. However in some very special situation it may be done.

ACKNOWLEDGEMENT:

This research was partially supported by Grant NSC81-0415-E-001-01 from the National Science Council of Taiwan, ROC.

REFERENCES

[1] M.M. Bugaighis, "Properties of the MLE for parameters of a Weibull regression model under type I censoring", *IEEE Transactions on Reliability*, Vol.39, 1990, pp 102-105.

[2] A.C. Cohen, "Tables for maximum likelihood estimatoes: Simply truncated and singly censored samples", *Technometrics*, Vol.3, 1961, pp 535-541.

[3] A.C. Cohen, "Maximum likelihood estimation in the Weibull distribution based on complete and censored samples", *Technometrics*, Vol.7, 1965, pp 579-588.

[4] D.R. Cox, "Regression models and life tables (with Discussion)", *Journal of the Royal Statistical Society, Series B*, Vol.34, 1972, pp 187-220.

[5] E.L. Crow, K. Shimizu, *Lognormal Distributions*, 1984; Marcel Dekker, New York.

[6] J. Deshayes, D. Picard, *Rupture de model'es en statistique*, 1988; These d'Etat, Orsay.

[7] T.B. Fomby, R.C. Hill, S.R. Johnson, *Advanced Econometrics Methods*, 1984; Academi Press, New York.

[8] D.V. Hinkley, "Inference about the intersection in two-phase regression", *Biometrika*, 56, 1969, pp 495-504.

[9] D.V. Hinkley, "Inference in two-phase regression", *Journal of the American Statistical Association*, Vol. 66, No.336, 1971, pp 736-743.

[10] V.K. Jandhyala, I.B. Macneill, "Tests of parameter changes at unknown times in linear regression models", *Journal of Statistical Planning and Inference*, 27, 1991, pp 291-316.

[11] J.F. Lawless, *Statistical Models and Methods for Lifetime Data*, 1982; John Wiley, New York.

[12] K.Y. Liang, S.G. Self and X. Liu, "The Cox proportional hazards model with change point: An epidemiologic application", *Biometrics*, Vol.46, 1990, pp 783-793.

[13] D.E. Matthews and V.T. Farewell, "On testing for a constant hazard against a change-point alternative", *Biometrics*, Vol.38, 1982, pp 463-468.

[14] W.Q.Jr. Meeker, W. Nelson, "Weibull variance and confidence limits by maximum likelihood for singly censored data", *Technometrics*, Vol.19, 1977, pp 473-476.

[15] W. Nelson, "Accelerated life testing – step-stress models and data analyses", *IEEE Transactions of Reliability*, Vol. R-29, No.2, 1980, pp 103-108.

[16] W. Nelson, *Accelerated Testing Statistical Models, Test Plans and Data Analyses*, 1990; John Wiley, New York.

[17] J.Schmee, G.L. Hahn, "A simple method for regression analysis with censored data", *Technomerics*, Vol.21, 1979, pp 417-432.

[18] K.J. Worsley, "Testing for a two-phase multiple regression", *Technometrics*, Vol.25, No.1, 1983, pp 35-42.

Accurate Inferences for the Reliability Function Considering Accelerated Life Tests

JORGE ALBERTO ACHCAR and JOSE CARLOS FOGO

Universidade de São Paulo, ICMSC, Caixa Postal 668,
13560-970, São Carlos, S. P., Brazil
and
Universidade Federal de São Carlos, DEs, Caixa Postal 676
13565-905, São Carlos, S. P., Brazil.

Keywords:
Accelerated life tests, reliability function, reparametrization.

Abstract:
With accelerated life tests assuming the power rule model and a Weibull distribution for the lifetimes of the units, we consider some aspects as accurate inferences on the reliability function at time t_0 in a specified stress level V_p. We use a reparametrization proposed by Guerrero and Johnson (1982) exploring a nonnormality measure for likelihood functions and posterior densities introduced by Kass and Slate (1992).

1. Introduction

Accelerated life tests are used in industrial applications when product life under normal operating conditions is very long but we need to obtain measures of reliability of units under the usual stress level (see for example, Mann, Schafer

*Kitsos, C.P. and Edler, L., Eds.:Industrial Statistics, Physica-Verlag, Heidelberg, 1997

and Singpurwalla, 1974; Nelson, 1990). Usually, statisticians consider the asymptotical normality to get inferences for the reliability function at time t_0 assuming different parametric models and censored observations (see for example, Lawless, 1982). These asymptotical results could be very poor, considering small or moderate sample sizes.

To improve these approximate inference results, we could consider an appropriate reparametrization to get good normality of the likelihood function (see for example, Anscombe, 1964; or Sprott, 1973, 1980).

Good parametrization also is very important to get accurate Bayesian inferences considering numerical or approximation methods for posterior moments or posterior densities of interest (see for example, Achcar and Smith, 1990; Kass and Slate, 1992; or Hills and Smith, 1993).

Consider T a random variable denoting the life time of a unit with a Weibull density,

$$f(t;\alpha,\beta)=\frac{\beta}{\alpha}\left(\frac{t}{\alpha}\right)^{\beta-1} exp\left\{-\left(\frac{t}{\alpha}\right)^{\beta}\right\} \tag{1}$$

where $t \geq 0$; $\alpha,\beta > 0$ and assume a stress variable V affecting the scale parameter α, but with common shape parameter β for all stress levels.

With k levels of a stress variable V, assume the power rule model,

$$\alpha_i = \frac{\delta}{V_i^{\gamma}} \tag{2}$$

where $i = 1, 2, \ldots, k$ and δ and γ are unknown parameters.

Also, assume a type II censoring mechanism, that is, the experiment terminates when we observe d_i failures for each stress level V_i, we have the ordered uncensored observations given by $t_{i(1)}, t_{i(2)}, \ldots, t_{i(d_i)}$ and $n_i - d_i$ censored observations equal to $t_{i(d_i)}$, $i = 1, 2, \ldots, k$.

Considering the data of k stress levels $V_1, V_2, \ldots, V_k$ taken at random, the likelihood function for β, γ and δ is given by

$$L(\beta,\gamma,\delta)=\frac{\beta^{d}}{\delta^{d\beta}}\left(\prod_{i=1}^{k} V_i^{d_i\beta\gamma}\right)\left(\prod_{i=1}^{k}\prod_{j\in D_i} t_{i(j)}^{\beta-1}\right) exp\left\{-\sum_{i=1}^{k}\frac{T_i(\beta)V_i^{\gamma\beta}}{\delta^{\beta}}\right\} \tag{3}$$

where $d=\sum_{i=1}^{k} d_i$, D_i is the set of uncensored observations at stress level V_i and $T_i(\beta)=\sum_{j=1}^{d_i} t_{i(j)}^{\beta}+(n_i-d_i)t_{i(d_i)}^{\beta}$.

Considering a specified stress level V_p, the reliability function at time t_0 is given by

$$R_p(t_0)=exp\left\{-\left(\frac{t_0}{\alpha_p}\right)^{\beta}\right\}. \tag{4}$$

From (2) and (4), we have $\delta=t_0\,V_p^{\gamma}\left(-ln\,r_p\right)^{-1/\beta}$, $0\le r_p=R_p(t_0)\le 1$, and the logarithm of the likelihood function for r_p, β and γ is given by

$$l(r_p,\gamma,\beta)=d\,ln\,\beta-d\,\beta\,ln\,t_0+\beta\gamma\,\mathbf{K}_1+(\beta-1)\mathbf{K}_2+d\,ln(-ln\,r_p)+$$

$$+(ln\,r_p)\sum_{i=1}^{k}T_i(\beta)\left[\frac{Q_i(\gamma)}{t_0}\right]^{\beta}, \tag{5}$$

where $\mathbf{K}_1=\sum_{i=1}^{k} d_i\ln V_i-d\ln V_p$, $\mathbf{K}_2=\sum_{i=1}^{k}\sum_{j\in D_i} ln\,t_{i(j)}$ and $Q_i(\gamma)=(V_i/V_p)^{\gamma}$.

Standard classical methods for inferences on r_p, β and γ usually requires the use of iterative methods to obtain the maximum likelihood estimators and the use of asymptotical results.

2. An useful reparametrization for the reliability function at time t_0

One way to improve the normality of the maximum likelihood estimator for the reliability function $R_p(t_0)$ at time t_0 and in a specified stress level V_p, is to consider different parametrizations or transformations of $R_p(t_0)$.

For a given λ, we consider a modified form of Guerrero and Johnson (1982) transformation given by

$$\phi_{GJ}(\lambda)=\left(\frac{R_p}{1-R_p}\right)^{\lambda}-1 \tag{6}$$

The transformation (6) has inverse given by

$$R_p = \frac{(\phi_{GJ}+1)^{1/\lambda}}{1+(\phi_{GJ}+1)^{1/\lambda}} \tag{7}$$

To consider the reparametrization (6), we should have an appropriate value of λ that gives good normality for the likelihood function of ϕ_{GJ}. One way to find this value assuming β and γ known, is to choose λ in (6) that gives third derivatives of the logarithm of the likelihood function $l[\phi_{GJ}(\lambda)]$ at the maximum likelihood estimator $\hat{\phi}_{GJ}(\lambda)$ in a standardised form,

$$STD[\hat{\phi}_{GJ}(\lambda)] = \left| l'''(\hat{\phi}_{GJ}(\lambda))\left[-l''(\hat{\phi}_{GJ}(\lambda))\right]^{-3/2}\right| \tag{8}$$

close to zero (see for example, Sprott, 1973; or Kass and Slate, 1992).
When all parameters are unknown, we could search for a joint transformation of $R_p(t_0)$, β and γ that gives joint normality for the likelihood function. One way to find this transformation is to explore some third derivative summaries that generalises the standardised third derivative (8) used in the one-parameter case (see Kass and Slate, 1992).
One of these measures is given by

$$m^2\overline{B}^2 = \sum_{a,b,c,d,e,f} d_{abc}\, d_{def}\, b_{ab}\, b_{de}\, b_{cf} \tag{9}$$

where b_{ab} are the elements of the inverse of the information matrix, d_{abc} denotes the third derivatives of the logarithm of the likelihood function locally at the maximum likelihood estimators and m is the number of parameters .
An alternative way is to search for an appropriate value of λ in reparametrization (6) for the reliability function $R_p(t_0)$ that gives close normality for the profile likelihood function of $\phi_{GJ}(\lambda)$.

3. The exponential case

Assuming $\beta = 1$ and the reparametrization (6) for the reliability function $R_p(t_0)$ at time t_0 and a stress level V_p, the logarithm of the likelihood function for ϕ_{GJ} and γ is given (from (5)) by

$$l(\phi_{GJ},\gamma) = -d\,ln\,t_0 + \gamma\,\mathbf{K}_1 + d\,ln\left[B(\phi_{GJ})\right] - \frac{B(\phi_{GJ})}{t_0}\sum_{i=1}^{k} T_i\,Q_i(\gamma) \qquad (10)$$

where $\mathbf{K}_1$ and $Q_i(\gamma)$ are defined in (5), $T_i = \sum_{j=1}^{d_i} t_{i(j)} + (n_i - d_i)t_{i(d_i)}$ and $B(\phi_{GJ}) = ln\left[1 + (1+\phi_{GJ})^{-1/\lambda}\right]$.

Assuming γ unknown, we should search for a appropriate value of λ in parametrization (6) such that

$$STD(\hat{\phi}_{GJ}) = \left| d^{-1/2}\left(2 - \frac{3B''(\hat{\phi}_{GJ})B(\hat{\phi}_{GJ})}{\left[B'(\hat{\phi}_{GJ})\right]^2}\right)\right| \qquad (11)$$

is close to zero ($\hat{\phi}_{GJ}$ maximises the profile likelihood function for ϕ_{GJ}).

Assuming γ known, we obtain from $STD(\hat{\phi}_{GJ}) = 0$

$\left(\text{where } \hat{\phi}_{GJ} = (\hat{r}_p^{-1} - 1)^{-\lambda} - 1;\ \hat{r}_p = exp\left\{-dt_0 \Big/ \sum_{i=1}^{k} T_i\,Q_i(\gamma)\right\}\right)$ the appropriate value of λ given by

$$\lambda = \left(\frac{2T}{3d\,t_0} + 1\right)\left(1 - e^{-d\,t_0/T}\right) - 1 \qquad (12)$$

where $T = \sum_{i-1}^{k} T_i Q_i$, $T_i = \sum_{j=1}^{d_i} t_{i(j)} + (n_i - d_i)t_{i(d_i)}$ and $Q_i = V_i^{\gamma}/V_p^{\gamma}$.

4. An example

In table 1, we have a generated type II censoring data set considering the power rule model (2) with $\gamma = 0.8$, $\delta = 500$ and an exponential distribution $(\beta = 1 \text{ in } (1))$.

From table 1 we have $d = \sum_{i=1}^{5} d_i = 65$.

The maximum likelihood estimators for γ and δ are given by $\hat{\gamma} = 0.7995$ and $\hat{\delta} = 500.5555$. The maximum likelihood estimator for the reliability function $R_1(t_0)$ at time $t_0 = 200$ considering the usual stress level $V_1 = 10$ is given by $\hat{r}_1 = 0.0806$.

Table 1: Generated data set with $\gamma = 0.8$ and $\delta = 500$.

i	V_i	n_i	d_i	$t_{i(j)}$
1	10	30	5	6, 8. 10, 12, 14
2	20	30	8	4, 5. 5, 6, 8, 8, 9, 14
3	30	30	12	2, 3. 3, 5, 6, 7, 7, 8, 8, 9, 10, 17
4	40	30	18	3, 3, 4, 5, 6, 6, 8, 9, 10, 10, 12,12, 13, 14, 14, 15, 24
5	50	30	22	2, 3, 4, 5, 5, 8, 8, 8, 9, 10, 12, 13,14, 14, 15, 18, 18, 18, 19, 20, 20,27

To improve the joint normality of the maximum likelihood estimators, we consider the transformation (from (6)) $\phi_{GJ}(\lambda)$ and γ where the appropriate value of λ minimises the third derivative summary (9). In table 2, we have the values for (9) considering $t_0 = 10$, 100, 200 and 400 in both parametrizations $(R_1(t_0), \gamma)$ and $(\phi_{GJ}(\lambda), \gamma)$.

Table 2: Values for $m^2\overline{B}^2$ (see (9)) and λ $(V_1 = 10)$.

t_0	$\hat{R}_1(t_0)$	$m^2\overline{B}^2(\hat{r}_1, \hat{\gamma})$	$\hat{\phi}_{GJ}(\lambda)$	$m^2\overline{B}^2(\hat{\phi}_{GJ}, \hat{\gamma})$	λ
10	0.8817	0.8589	0.1918	0.0014	0.0874
100	0.3201	0.4413	-0.2563	0.0014	0.3201
200	0.0806	5.9022	-0.4739	0.0014	0.2639
400	0.0065	35.5343	-0.5263	0.0014	0.1486

In figures 1 and 2, we have contour plots for the likelihood functions in both parametrizations. We observe a good improvement in the normality of the likelihood function for $\phi_{GJ}(\lambda)$ and γ, specially for $t_0 = 400$, since we have elliptical form for the contour plot (see figure 2).

We also could check the normality of the profile likelihood function in parametrization $\phi_{GJ}(\lambda)$ considering the t-plot (see Hills and Smith, 1993) $T(\phi_{GJ})$ against some values of ϕ_{GJ}, where,

$$T(\phi_{GJ}) = \text{sgn}(\phi_{GJ} - \hat{\phi}_{GJ})\{-2l_{\hat{\gamma}}(\phi_{GJ}) + 2l_{\hat{\gamma}}(\hat{\phi}_{GJ})\}^{1/2} \quad (13)$$

$\hat{\phi}_{GJ}$ is the maximum of the profile likelihood function $L_{\hat{\gamma}}(\phi_{GJ}(\lambda))$, with the appropriate values of λ given by (8). Since we observe straight lines (see figure 3), we conclude by the normality of the profile likelihood function for $\phi_{GJ}(\lambda)$. In the original parametrization, the plots of $T(R_1(t_0))$ against $R_1(t_0)$ are markedly curved (see figure 4), which indicates the nonnormality of the profile likelihood function for $R_1(t_0)$, with $t_0 = 10$, 200 and 400.

Figure 1: Contour plots for the likelihood function $L(r_1, \gamma)$ $(V_1 = 10)$.

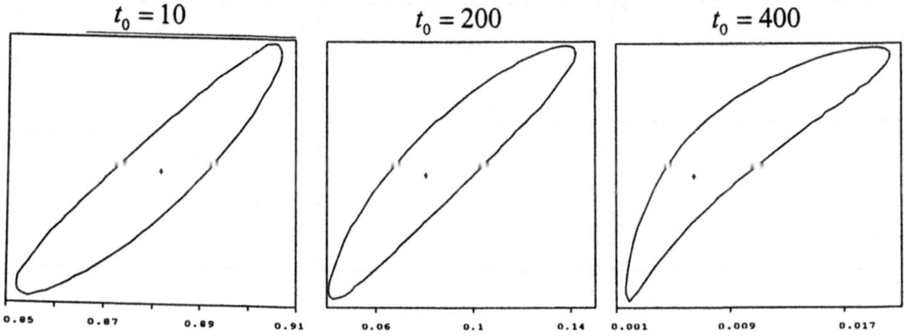

Assuming $\hat{\gamma} = 0.7995$ known, we have in table 3 approximate 95% confidence intervals for $R_1(t_0)$ with $t_0 = 10$, 100, 200 and 400 considering the asymptotical normality for the maximum likelihood estimators in both parametrizations $R_1(t_0)$ and $\phi_{GJ}(\lambda)$ and exact 95% confidence intervals for $R_1(t_0)$ based on the fact that $2\,T(-\ln r_1)/t_0$ has an exact chi-square distribution with $2d$ degrees of freedom. We observe good inference results considering the parametrization $\phi_{GJ}(\lambda)$ with the appropriate values of λ, especially for extreme values of t_0.

Figure 2: Contour plots for the likelihood function $L(\phi_{GJ},\gamma)$ $(V_1 = 10)$.

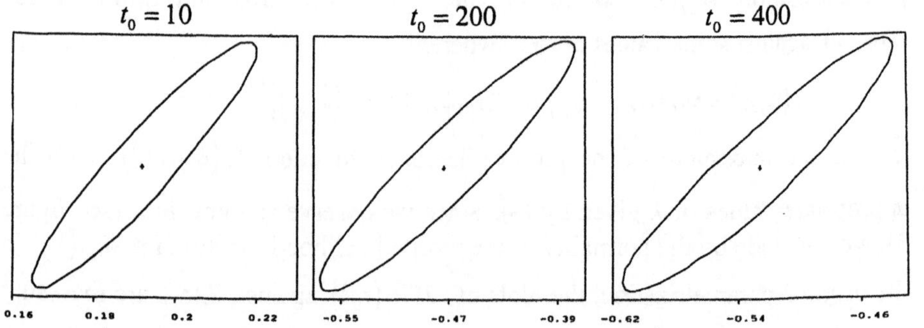

Figure 3: T-plot of the profile likelihood function $L_{\hat{\gamma}}(\phi_{GJ}(\lambda))$ $(V_1 = 10)$.

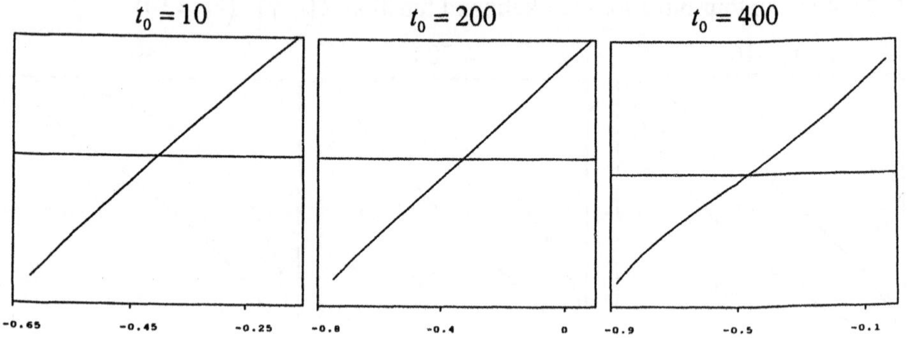

Figure 4: T-plot of the profile likelihood function $L_{\hat{\gamma}}(r_1)$ $(V_1 = 10)$.

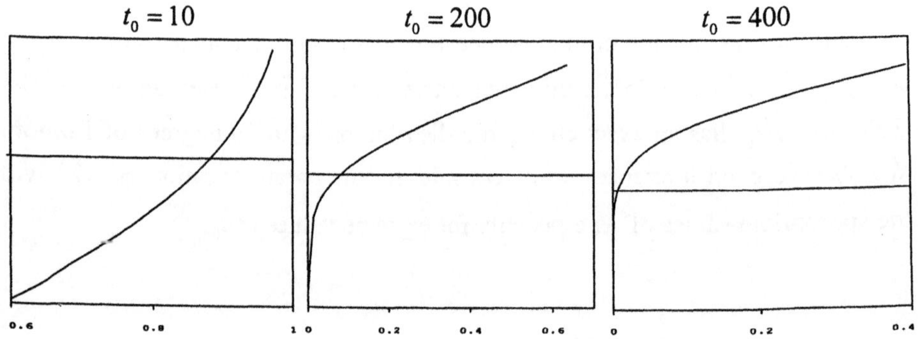

In table 4, we have Laplace's approximations (see for example, Tierney and Kadane, 1986) for the posterior mean $E(R_1(t_0)/\text{data})$ with $\beta = 1$ and $\gamma = 0.7995$ known, considering a Jeffreys noninformative prior density for $R_1(t_0)$, $\pi_0(r_1/\text{dados}) \propto 1/(-\ln r_1)\, r_1$, $0 \le r_1 \le 1$ (see for example, Martz and Waller, 1982). Since this case, we have an exact expression for $E(R_1(t_0)/\text{data})$, we observe very accurate approximate Bayesian inferences for $R_1(t_0)$ considering the parametrization $\phi_{GJ}(\lambda)$ with λ given by (12).

Table 3: 95% confidence intervals for $R_1(t_0)$ $(V_1 = 10)$.

t_0	Exact distribution for $2T(-\ln r_1)/t_0$	Asymptotical normality for $\hat{R}_1(t_0)$	λ given by (12)	Asymptotical normality for $\hat{\phi}_{GJ}(\lambda)$
10	(0.8536 , 0.9074)	(0.8547 , 0.9086)	-0.2553	(0.8530 , 0.9069)
100	(0.2054 , 0.3761)	(0.1971 , 0.3709)	0.0952	(0.2040 , 0.3761)
200	(0.0422 , 0.1433)	(0.0313 , 0.1300)	0.1628	(0.0415 , 0.1416)
400	(0.0018 , 0.0205)	(-0.0015 , 0.0145)	0.1250	(0.0017 , 0.0202)

5. Concluding remarks

The use of transformation (6) for the reliability function at a specified time t_0 and in a fixed stress level V_p considering an appropriate value for λ obtained from the standardised third derivative summaries (8) or (9) could be of great practical interest to improve the accuracy of the asymptotical inferences.

Table 4: Posterior means for $R_1(t_0)$ $(V_1 = 10)$.

t_0	Exact	Laplace's Approximations			
		Reparametrization $R_1(t_0)$	Percentage errors	Reparametrization $\phi_{GJ}(\lambda)$	Percentage errors
10	0.8818	0.8818	(0.000)	0.8818	(0.000)
100	0.2874	0.2874	(0.009)	0.2874	(0.002)
200	0.0846	0.0848	(0.203)	0.0845	(0.018)
400	0.0078	0.0080	(2.540)	0.0078	(0.052)

As it was observed in this paper, the invertible reparametrization (6) is easily obtained for each application considering accelerated life test models. We also could extend these results to other stress-response models used in engineering applications, like the Arrhenius or Eyring models (see for example, Nelson, 1990), and to consider other parametrical distributions for the lifetimes of the units.

References

ACHCAR, J. A.; SMITH, A. F. M. (1990) - Aspects of Reparametrization in Approximate Bayesian Inference. In Essays in Honour of G. A. Barnard. - Ed. J.Hodges, pp. 439-452 - Amsterdam, North Holland.

ANSCOMBE, F. J. (1964) - Normal Likelihood Functions. - Ann. Inst. Stat. Math., 16, 1-19.

GUERRERO, V. M.; JOHNSON, R. A. (1982) - Use of the Box-Cox Transformation with Binary Response Models. - Biometrika, 69, 309-314.

HILLS, S. E.; SMITH, A. F. M. (1993) - Diagnostic Plots for Improved Parametrization in Bayesian Inference. - Biometrika, 80, 61-74.

KASS, R. E.; SLATE, E. H. (1992) - Reparametrization and Diagnostics of Posterior Nonnormality. - In Bayesian Statistics 4, Ed. J.M. Bernardo, J.O. Berger, A.P. Dawid and A.F.M. Smith, pp. 289-306 - Oxford University Press.

LAWLESS, J. F. (1982) - Statistical Model and Methods for Lifetime Data. - New York, John Wiley & Sons.

MANN, N. R.; SCHAFER, R. E.; SINGPURWALLA, N. D. (1974) - Methods for Statistical Analysis of Reliability and Life Data. - New York, John Wiley & Sons.

MARTZ, H. F.; WALLER, R. A. (1982) - Bayesian Reliability Analysis. - New York, John Wiley & Sons.

NELSON, W. (1990) - Accelerated Testing: Statistical Models, Test Plans and Data Analysis. - New York, John Wiley & Sons.

SPROTT, D. A. (1973) - Normal Likelihoods and Their Relation to Large Sample Theory of Estimation. - Biometrika, 60,3, 457-465.

SPROTT, D. A. (1980) - Maximum Likelihood in Small Samples: Estimation in the Presence of Nuisance Parameters. - Biometrika, 67, 515-523.

TIERNEY, L. E KADANE , J. B. (1986) - Accurate Approximations for Posterior Moments and Marginal Densities. - JASA, 81, 82-86.

Prediction of Failures that Have Never Occured: Exponential Case

JAROMIR ANTOCH AND JOSEF MACHEK
Charles University, Department of Statistics
Sokolovská 83, CZ – 186 00 Praha 8, Czech Republic

Abstract

Failures of some components, rotors of turbines or critical parts of aircraft etc., belong to extremely rare events even if respective components have been in service for a long time. Sometimes it even happens, that components of given type are observed for years (several decades) and no failures have occurred all the while. In this paper we shall describe several possibilities what to do in such a situation under the assumption, that the stream of failures form Poisson process. Advantages and disadvantages of MLE, UMVE and Bayesian estimators are covered and thoroughly discussed.

Key Words: Availability, Bayesian estimator, confidence limits, MLE, UMVE.

1 Problem

The topic partly covered in this paper steamed out from the analysis of the lifetimes of turbine rotors produced by SKODA a.s., Pilsen, Czech Republic. There are several hundreds of them in service all over the world, some for more than thirty years. However, the only information at the end of the year 1996 was that there were no failures observed during that time period. It is evident that the natural question *What conclusions, if any, can be drawn from such a data?* arises. In other words, how to estimate (reliability) parameters if all the observed (life)times are smaller than the time to the first failure.

Due to the limited space available, this paper covers only the situation when we assume that times to failure are exponentially distributed. The analogous situation for the Weibull distributed data is covered in Antoch et al. (1997).

*Kitsos, C.P. and Edler, L., Eds.:Industrial Statistics, Physica-Verlag, Heidelberg, 1997

2 Statistical model and criteria of reliability used

Let the time to failure of given type for some component be a random variable X with the distribution function $F(x) = 0$ for $x \leq 0$, and $F(x)$ continuous and strictly increasing for $x > 0$. Moreover, let in the case of failure the component be immediately repaired in such a way that it can be considered as new. From the point of view of the theory of reliability it means, that we assume simple renewal process. Thus, the behavior of each component during time period $(0, t]$ can be described either by the times between failures $X_1, X_2, \ldots$, or using the times $S_1, S_2, \ldots$ in which the failures appeared, i.e. $S_1 = X_1, S_2 = X_1 + X_2, \ldots, S_k = X_1 + \ldots + X_k$ etc. Due to the assumption that after each reparation the object is in the same state as being new, the random variables $X_1, X_2, \ldots$ are independent and identically distributed (iid) with the same distribution function $F(x)$ and density $f(x)$.

In this paper we shall deal with the following characteristics of reliability:

1) *Mean (expected) time between failures*

$$EX = \int_0^\infty x f(x)\, dx = \int_0^\infty x\, dF(x).$$

2) *Mean (expected) number of failures in time period* $(0, t]$

$$EN(t) = \sum_{k=1}^{\infty} k \times P\big(N(t) = k\big)$$

where $N(t)$ denotes the number of failures in time period $(0, t]$.

3) *Probability of the service without failures in time period* $(0, t]$

$$Q(t) \equiv P\big(N(t) = 0\big) = P\big(X_1 > t\big) = 1 - F(t).$$

4) *Intensity of failures*

$$r(t) = \frac{f(t)}{1 - F(t)}, \qquad t > 0.$$

All the above characteristics can be derived from both $F(x)$ and $f(x)$. Moreover, concerning the $EN(t)$, following bounds can be often useful

$$F(t) \leq EN(t) \leq \frac{F(t)}{1 - F(t)}.$$

3 Model and basic estimators

In this section we shall assume that the time between failures X follows exponential distribution with

$$f(x) = \begin{cases} \lambda e^{-\lambda x}, \\ 0, \end{cases} \quad \text{and} \quad F(x) = \begin{cases} 1 - e^{-\lambda x}, & x > 0,\ \lambda > 0, \\ 0, & \text{otherwise.} \end{cases}$$

Parameter λ denotes so called *intensity of failures*, and the process of failures in this case is *Poisson process.*

It is evident, that the characteristics of reliability described above can be expressed in terms of λ as follows:

- Mean time to failure $EX = 1/\lambda$.
- Expected number of failures in time period $(0, t]$ $EN(t) = \lambda t$.
- Probability of the service without failures in time period $(0, t]$ $Q(t) = P(X > t) = e^{-\lambda t}$.

When estimating parameter λ from real data, we can sometimes meet quite serious problem, i.e. that the number of failures is very small or even equal to zero. This means, that we can state only the fact $X_i > t_i$, where t_i is a given number describing the time in service of the ith observed component, and that during this time no failures were encountered.

More precisely, let us observe n components, ith of them during time period $(0, t_i]$, where t_i corresponds to the time from launching it to the service till the time of the analysis (end of the observation). Moreover, let ith component has during this time r_i failures. Under these assumptions the numbers of failures are realizations of mutually independent random variables R_i, each of them following Poisson distribution with

$$P(R_i = r_i) = \frac{(\lambda t_i)^{r_i} e^{-\lambda t_i}}{r_i!}, \qquad r_i = 0, 1, 2, \ldots$$

Joint distribution of all R_i's is

$$P(R_1 = r_1, \ldots, R_n = r_n) = \lambda^{\sum_{i=1}^{n} r_i} \exp\left\{-\lambda \sum_{i=1}^{n} t_i\right\} \prod_{i=1}^{n} \frac{t_i^{r_i}}{r_i!}. \tag{1}$$

It is evident that $R = \sum_{i=1}^{n} R_i$ is a sufficient statistics for the parameter λ. Moreover, random variable R, i.e. the overall number of failures observed during time period $t = \sum_{i=1}^{n} t_i$ (total time in service) has Poisson distribution

$$P(R = r) = \frac{(\lambda t)^r e^{-\lambda t}}{r!}, \qquad r = 0, 1, 2, \ldots$$

Therefore, the mean value of R is $ER = \lambda t$ and the estimator

$$\hat{\lambda} = \frac{R}{t} = \frac{\sum_{i=1}^{n} R_i}{\sum_{i=1}^{n} t_i} = \frac{\text{total number of failures}}{\text{total time in service}} \tag{2}$$

The real problem arises when $R = 0$, being quite probable situation. The consequent use of the method of maximum likelihood suggests to estimate λ by zero, i.e. to put

$$\widehat{\lambda} = 0 \qquad if \qquad R = 0.$$

Indeed, likelihood function in such a case attains its maximum at the point $\lambda = 0$.

On the other hand, the mean number of failures $N(t)$ in time period $(0,t]$ is λt and the probability of at least one failure during time period $(0,t]$ is $1 - \exp\{-\lambda t\}$. Therefore, for $\lambda = 0$ and arbitrary large t the mean number of errors $EN(t) = 0$ and the probability of appearance of at least one error is equal to $1 - Q(t) = 0$. However, *such an estimator contradicts to the common sense and is practically unacceptable.* The fact, that we observed no failure during time period of any length gives us no right for a conclusion, that the failure will never occur in the future. Therefore, in the following sections we shall describe several possibilities what to do in such a situation.

4 Possible modifications of $\widehat{\lambda}$

First idea, that comes into ones mind, is to modify "slightly" but "reasonably" the formula for the maximum likelihood estimator provided $R = 0$.

One possibility is to assume in such a case that one failure appeared just at the end of the observation, i.e. *to work with some kind of one fictive failure.* Modified maximum likelihood estimator in such a case will be

$$\widehat{\lambda}_{m_1} = \begin{cases} 1/t, & R = 0, \\ R/t, & R = 1, 2, \ldots \end{cases}$$

When we try to describe the behavior of $\widehat{\lambda}_{m_1}$ using classical criteria, especially the unbiasedness, we can easily show that

$$E\,\widehat{\lambda}_{m_1} = \frac{e^{-\lambda t}}{t} + \sum_{r=1}^{\infty} \frac{r}{t} \times \frac{e^{-\lambda t}(\lambda t)^r}{r!} = \lambda + \frac{e^{-\lambda t}}{t}.$$

This means that $\widehat{\lambda}_{m_1}$ is biased to the right, i.e. it is conservative, and the real value of λ is systematically overestimated by

$$\frac{e^{-\lambda t}}{t} = \exp\left\{-\lambda \sum_{i=1}^{n} t_i\right\} \Big/ \sum_{i=1}^{n} t_i.$$

Moreover, it is not natural to have the same estimate for both $r = 0$ and $r = 1$.

Some statisticians tried to decrease the bias of $\widehat{\lambda}_{m_1}$ using another modification, i.e.

$$\widehat{\lambda}_{m_2} = \begin{cases} 1/3t, & R = 0, \\ R/t, & R = 1, 2, \ldots \end{cases}$$

On one hand the bias will decrease. Mechanical use of such an estimator, however, can result in a serious underestimation of the real value of λ.

Assume for example, that the total time in service is t. If $\lambda = 1/t$, then $P(R = 0) = \exp\{-\lambda t\} = 1/e \approx 0.37$. However, it is evident that in such a situation the estimator $\widehat{\lambda}_{m_2} = 1/3t$ drastically underestimates the 'true value' $\lambda = 1/t$. Therefore, we suggest the use of another modifications supported by the arguments from following section.

5 Bayesian approach

When applying Bayes method, we assume that parameter λ is a realization of random variable Λ with some distribution, so called *prior distribution* (of the parameter of interest). This distribution must be known before the experiment. Distribution of the result of experiment, i.e. distribution of the number R of failures observed during time period $(0, t]$, which is given in our case by

$$P(R = r) = \frac{(\lambda t)^r e^{-\lambda t}}{r!}, \qquad r = 0, 1, \ldots 3$$

is considered to be the conditional distribution of R for given value of (the random parameter) λ. For fixed result of the experiment (in our case $R = r$) we can calculate according to the Bayes theorem density of the variable Λ conditioned by the result of the experiment, i.e. so called *posterior distribution* $h(\lambda \mid R = r)$. As the estimator of the parameter λ one typically uses either mean, modus or median of the posterior distribution.

The result is up to some extent (sometimes heavily) dependent on the choice of the prior distribution. In the case when we have almost no information about values of the parameter and corresponding probabilities, we usually follow the *principle of uncertainty* and use the distribution expressing absolute lack of knowledge. In other words, we use as the prior distribution of (the random parameter) λ uniform distribution on some interval $(0, L)$. This means, that any value λ between 0 and L can appear quite deliberately and the value $\lambda > L$ is practically impossible. Formally, the prior distribution in this case is

$$g(\lambda) = \begin{cases} 1/\mathcal{L}, & 0 < \lambda \le \mathcal{L}, \\ 0, & \text{otherwise,} \end{cases}$$

so that the posterior density for given $R = r$ has the form

$$h(\lambda \mid R = r) = \frac{1}{L}\frac{(\lambda t)^r e^{-\lambda t}}{r!} \Big/ \int_0^L \frac{1}{L}\frac{(\lambda t)^r e^{-\lambda t}}{r!}\, d\lambda = \frac{(\lambda t)^r e^{-\lambda t}}{\int_0^L (\lambda t)^r e^{-\lambda t}\, d\lambda}\,.$$

It is easy to show that

$$\int_0^L (\lambda t)^r e^{-\lambda t}\, d\lambda = \frac{1}{t}\int_0^{Lt} v^r e^{-v}\, dv\,,$$

from where for $r = 0$ we get

$$\int_0^L e^{-\lambda t}\, d\lambda = \frac{1 - e^{-Lt}}{t}\,.$$

Thus, posterior distribution for $R = 0$ has density

$$h(\lambda \mid R = 0) = \begin{cases} \dfrac{te^{-\lambda t}}{1 - e^{-\mathcal{L}t}}\,, & 0 < \lambda \leq \mathcal{L}, \\ 0, & \text{otherwise.} \end{cases}$$

If we admit, moreover, that it is difficult to bound *a priori* possible values of λ by some number L, i.e. if we let $L \to \infty$, then the posterior density of λ for $R = 0$ is

$$h(\lambda \mid R = 0) = \begin{cases} te^{-\lambda t}, & 0 < \lambda < \infty, \\ 0, & \text{otherwise,} \end{cases}$$

being nothing else than the density of exponential distribution with the mean $1/t$.

Therefore, the *mean type Bayesian estimator* of λ in the case when $R = 0$ is given by

$$\widehat{\bar{\lambda}}_B = \frac{1}{t}, \qquad t > 0\,.$$

The letter B recalls that the estimator is of the Bayesian type, the bar over the intensity λ says that it is the Bayesian estimator derived as the mean of the posterior distribution and hat stands for the estimator. We can see immediately that $\widehat{\bar{\lambda}}_B$ is equivalent to the maximum likelihood estimator with "one fictive error", so that all objections stated above hold also for this estimator, what makes it practically unacceptable.

Let us notice, that modus of the posterior distribution of λ for $R = 0$ is equal to zero, so that the *modus type Bayesian estimator* in the case when $R = 0$ coincides with the maximum likelihood estimator, being thus also unusable.

Finally, the *median type Bayesian estimator* $\hat{\hat{\lambda}}_B$ is equal to the median of the posterior distribution, i.e. $\hat{\hat{\lambda}}_B$ must satisfy

$$P(\hat{\hat{\lambda}}_B \geq \lambda \mid R = 0) = P(\hat{\hat{\lambda}}_B \leq \lambda \mid R = 0) = 1/2,$$

i.e. it is the root of the equation

$$e^{-\lambda t} = 1/2$$

from where

$$\hat{\hat{\lambda}}_B = \frac{\log 2}{t} = \frac{1}{1.443t}, \qquad t > 0.4$$

This estimator seems to be quite reasonable. The results of the following section, however, show that it is still far away from an "optimal" one.

Aside "uniform prior" we considered also several other possibilities. Interesting and immediately applicable result was obtained when we used so called *Jeffreys' prior density* given by

$$g^G(\lambda) = C\sqrt{I(\lambda)}$$

where $I(\lambda)$ denotes *Fisher information about parameter* λ contained in the realization of the process of failures of the length t and C is normalizing constant. In our case the Fisher information is equal to

$$I(\lambda) = -E\left\{\frac{\partial^2 \log f(r)}{\partial \lambda^2}\right\} = -E\left\{-\frac{r}{\lambda^2}\right\} = \frac{t}{\lambda}$$

where $f(r)$ denotes density of the Poisson distribution with parameter λ. Applying all our assumptions (and above mentioned facts), we arrive to the posterior density

$$h^G(\lambda \mid r, t) = e^{-\lambda t}(\lambda t)^r \sqrt{\frac{t}{\lambda}} \Big/ \int_0^\infty e^{-\lambda t}(\lambda t)^r \sqrt{\frac{t}{\lambda}}\, d\lambda = \frac{e^{-\lambda t}(\lambda t)^{r-1/2} t}{\Gamma(r + 1/2)}$$

where

$$\Gamma(r + 1/2) = \int_0^\infty e^{-v} v^{r+1/2-1}\, dv = \int_0^\infty e^{-v} v^{r-1/2}\, dv\,.$$

This means that the *posterior distribution of the parameter* λ for given number of failures $R = r$ during the time period $(0, t]$ is of gamma type with the shape parameter equal to $r + 1/2$. Mean value of this distribution is $(r + 1/2)/t$, from where the *mean type Bayesian estimator* of λ, provided $R = 0$ and Jeffreys' prior density is used, has the form

$$\hat{\lambda}_B^G = \frac{1}{2t}, \qquad t > 0,$$

being the estimator quite realistic and satisfying all natural intuitive requirements.

Concerning the *modus type Bayesian estimators* based on the Jeffreys' prior density, its value for $r = 0$ is zero and has no sense.

Finally, it can be shown (analogously as above) that the *median type Bayesian estimator* based on the Jeffreys' prior density has for $R = 0$ the form

$$\hat{\lambda}_B^G = \frac{1}{4.396\,t}, \qquad t > 0.$$

However, it is evident that this estimator underestimates the real value of λ too much, so that we cannot suggest it for the practical use.

Aside the uniform prior distribution representing the total uncertainty, resp. Jeffreys' prior density which offers us the independence of the posterior distribution on the initial parametrization of the model, it seems to be quite natural to use conjugate system of prior distributions. It is well known that in our situation, where we assume that the distribution of the result of experiment (i.e. distribution of the number R of failures observed during time period $(0, t]$) is given by (3), the natural system of conjugate distributions is formed by the family of gamma distributions. More precisely, if the prior density follows $\Gamma(a, p)$, then after straightforward calculation we can easily show that the posterior distribution follows $\Gamma(a+t, p+r)$ and the *mean type Bayesian estimator* of λ is of the form

$$\hat{\lambda}_B^C = \frac{p+r}{a+t} \cdot 5$$

Quite analogously the *median type Bayesian estimator* is equal (for p integer) to

$$\hat{\lambda}_B^C = \frac{\chi^2_{2(p+r),0.5}}{2(a+t)}.$$

If p is not integer, the estimator must be found numerically.

It is evident, however, that natural question *how to chose parameters* (a, p) *determining the prior distribution* arises. Interesting choice is $p = 1$, which implies that the prior distribution is of the exponential type. Therefore, let us take a look what happens if we change the value of the second parameter. We shall start with a small, more precisely, we let $a \to 0$. Then it is evident that

$$\lim_{a \to 0} \hat{\lambda}_B^C = \frac{1+r}{t}.$$

If we put, moreover, $r = 0$, being the case that corresponds to the situation of our main interest (i.e. when no failures were observed at all), resulting estimator is nothing else than the maximum likelihood estimator with one fictive error, resp. standard mean type Bayesian estimator provided $R = 0$. The opposite situation, i.e. $a \to \infty$, leads to unacceptable estimator of λ equal to 0.

Quite interesting situation arrives when we put $p = 1/2, r = 0$ and let $a \to 0$. Then the mean type Bayesian estimator has form $1/2t$. Quite analogously, for $p = 1/3, r = 0$ and $a \to 0$ the mean type Bayesian estimator is equal to $1/3t$ etc. Notice, that while the choice of "small" a expresses the conviction that the value of λ can be "large", the choice of $r = 1/3$ on the other hand indicates our "belief" that the value of λ will be "rather smaller than larger".

Another possibility is to chose p as a multiple of the parameter a, i.e. to put $p = ca$. Then the choice of a large enough leads to the concentration of (5) in the neighborhood of c, what corresponds to the "strong expectation" that the true value of the parameter λ will be close to c. Corresponding *mean type Bayes estimator* is of the form

$$\hat{\hat{\lambda}}_B^C = \frac{c + r/a}{1 + t/a}.$$

For large values of a both $r/a \approx 0$ and $t/a \approx 0$, so that the estimator itself practically does not depend on the overall time of the observation. On the contrary, the situation $a \to 0$ implies the mean type Bayesian estimator of the form r/t, being for us very well known situation thoroughly discussed above.

Summarizing all these remarks, it seems to us that in the case when we do not have really important reasons to assume that the value of the parameter λ is close to c, the most reasonable choice seems to be the use of $p = 1$ and $a \to 0$, what leads to the same result as the choice of the uniform prior with $L \to \infty$.

Finally, we have concentrated on the use of so called predictive densities. However, the results were so non-perspective, that we shall not discussed them here.

6 Neyman-Pearson's approach

Applying Neyman-Pearson's theory, upper confidence limit for unknown parameter with the confidence coefficient $1 - \alpha$ can be easily constructed which, using small enough value of α, can help us to bound "arbitrarily" the risk of underestimation of the true value of the parameter.

If the unknown parameter is intensity λ, and the result of the experiment is number of failures (of given type) during time period $(0, t]$, the upper confidence limit $\lambda_{U,1-\alpha}$ is given by

$$\lambda_{U,1-\alpha} = \frac{1}{2t}\chi^2_{2(r+1),1-\alpha}$$

where $\chi^2_{2(r+1),1-\alpha}$ denotes $100(1-\alpha)\%$ quantile of the χ^2 distribution with $2(r+1)$ degrees of freedom.

This results in the case $R = 0$, which is of our special interest, in

$$\lambda_{U,0.5} = \frac{\log 2}{t} = \frac{1}{1.443\,t}, \qquad t > 0.$$

Notice, that this coincides with the Bayesian estimator based on the median of the posterior distribution with uniform prior, see (4), being somehow unexpected result. In addition to other remarks stated above, this result suggests that the median based Bayesian estimator is not probably an excellent estimator.

7 Probability of the all year round service without failures

Let random variable X, representing the time to failure, follows exponential distribution with the mean value $1/\lambda$, i.e. let us observe Poisson process with intensity λ. From the previous sections it is evident that probability of no failure during time period $(0, t_0]$ is equal to

$$Q \equiv Q(t_0) = P(X > t_0) = e^{-\lambda t_0}.$$

Let us assume, analogously as above, that we observed n components, ith of them during time period $(0, t_i]$, $i = 1, \ldots, n$, so that the total time in service for all components under consideration is $t = \sum_{i=1}^{n} t_i$. Let us denote by R the number of failures of given type observed during this time period.

If $t_0 < t$, then the unbiased estimator with minimal variance of the probability Q is equal to

$$\widehat{Q} = \left(1 - \frac{t_0}{t}\right)^R, \qquad R = 0, 1, 2, \ldots$$

Another estimator that might be considered is the maximum likelihood estimator

$$\widehat{Q}_M = e^{-\widehat{\lambda} t_0},$$

where $\widehat{\lambda}$ is MLE of λ given by (2).

The greatest disadvantage of both these estimators is the fact, that for $R = 0$ we have

$$\widehat{Q} = \widehat{Q}_M = 1,$$

i.e. practically unacceptable result. Therefore, we suggest to use rather

$$\widehat{Q}_m = e^{-\widehat{\lambda}_m t_0} \approx 1 - \widehat{\lambda}_m t_0,$$

where $\widehat{\lambda}_m$ is either one of the properly modified maximum likelihood estimators or one of the Bayesian estimators of λ. Some possible modifications are described in preceeding sections.

Another possibility one can use for the estimation of Q is the lower confidence bound for the whole year round service without failure of given type, which is given by

$$Q_L = e^{-\lambda_{U,1-\alpha} t_0} \approx 1 - \lambda_{U,1-\alpha} t_0$$

with $\lambda_{U,1-\alpha}$ given by

$$\lambda_{U,1-\alpha} = \frac{1}{2t} \chi^2_{2(r+1),1-\alpha}.$$

Some other results concerning different aspects of the estimators of reliability function in the case of exponentially distributed data as well as comparison of respective methods can be found, e.g., in Antoch, Linka and Brzezina (1997), Harris and Soms (1988), Hurt (1976) or Martz and Waller (1982).

Acknowledgement

This research was partially supported by a grant GA ČR 201/97/1163.

References

[1] Antoch J., Klaschka J., Machek J. and Straka F., *Prediction of failures that have never occurred – Weibull case.* Proceedings of ASMDA'97. (Lauro N. C., ed.), 1997 (to appear).

[2] Antoch J., Linka A. and Brzezina M. *Asymptotic approximation of Bayes risk of estimators of reliability for exponentially distributed data.* Statistics & Decision (1997) (to appear).

[3] Antoch J. and Machek J. *Risk of appearance of turbine missiles.* Technical report, Charles University, Deparment of Statistics, Praha, 1996a, (in Czech).

[4] Antoch J. and Machek J. *Risk of the crash of turbine's rotor* Technical report, Charles University, Department of Statistics Praha 1996b, (in Czech).

[5] Harris B. and Soms A. *Sensitivity and asymptotic properties of Bayesian reliability estimates in the exponential case.* Statistics & Decision, 33 – 47 1988.

[6] Hurt J. *On estimation of reliability in exponential case.* Aplikace matematiky, 263–272, 1976.

[7] Jeffreys H. *Theory of Probability*, 3rd ed. Clarendon Press, Oxford, 1961.

[8] Jones S. P. *Prediction of failure events when no failure have occurred.* COMPSTAT'96, Physica-Verlag, Heidelberg. (Prat A., ed.), pages 307–312, 1996.

[9] Lehmann E. L. *Theory of Point Estimation*, 2nd ed. Wadsworth & Brooks/Cole, California, 1991.

[10] Martz H. F. and Waller R. A. *Bayesian Reliability Analysis.* J. Wiley, New York, 1982.

[11] Pugh E. L. *The best estimate of reliability in exponential case.* Oper. Res., **11/17**, 57–61, 1963.

[12] US Nuclear Regulatory Commission. *Protection against Low Trajectory Turbine Missiles.* Regulatory Guide 1.115, 1977.

[13] US Nuclear Regulatory Commission. *Safety Analysis Reports for Nuclear Power Plants.* Regulatory Guide 1.70, Revision 3, 1978.

[14] US Nuclear Regulatory Commission *Standard Review Plan.* NUREG-0800, 1981.

Algorithm of Confidence Limits Calculation for the Probability of the Value 1 of a Monotone Boolean Function of Random Variables

ALEXANDER M. ANDRONOV

Department of Control Computer Systems, Riga Aviation University,
1, Lomonosov Str., Riga, LV-1019 Latvia
e-mail: rau@rau.vernet.lv

Key words:
Boolean function, probability of the value 1, lower confidence limit, reduced gradient method.

Abstract:
Confidence limits techniques have been used for a long time in reliability problems. Various types of systems and statistical data have been considered in the literature. In this paper we discuss some general algorithm of confidence limits calculation which is suitable for all the monotone Boolean functions. This allows to find numerically confidence limits in various problems which arise in the reliability, the inventory, the system maintenance and others.

Glossary of Acronyms:
CDNF - Canonical disjunctive normal form.

1. Introduction

Boolean functions are a popular technique for modelling multicomponent systems of logical control, structural reliability, fault detection etc. Many of these models are probabilistical in nature which means that the arguments of Boolean functions are random variables. The random Boolean functions have

*Kitsos, C.P. and Edler, L., Eds.:Industrial Statistics, Physica-Verlag, Heidelberg, 1997

been considered often in the literature (Beichelt & Franken, 1983; Schneeweiss, 1989; Andronov, 1996). A classical statistical problem is to estimate sequential-parallel system reliability by binomial trials of its components. An important problems of an interval estimation have been considered in (Gnedenko, 1969, 1983; Mann et al., 1983). In this paper, we describe a general algorithm of the interval estimation of the probability of the value 1 for monotone Boolean functions of random variables.

Let $B = \{0,1\}$; $f(x_1, x_2, \ldots, x_k)$ be a monotone increasing boolean function $B^k \to B$; $X_1, X_2, \ldots, X_k$ represent mutually independent random boolean variables. The probabilities $p_i = P\{X_i = 1\}$ are unknown, but for $i = \overline{1,k}$ we have S_i "success" (unit value of X_i) from the n_i independent Bernoulli
trials. We denote $\mathrm{S} = (S_1, S_2, \ldots, S_k)$, $\mathrm{p} = (p_1, p_2, \ldots, p_k)$, $\theta_p = P_p\{f(x_1, x_2, \ldots, x_k) = 1\}$. The lower limits $\underline{\theta}(\mathrm{S})$ of the γ-confidence interval for the probability θ_p have to be found:

$$P_p\{\underline{\theta}(\mathrm{S}) \le \theta_p\} = \gamma .$$

A numerical method of confidence limits determination is described in Section 2. A corresponding algorithm is detailed in Section 3.

2. Method of confidence limits calculation

Our approach is based on the concept of confidence sets construction, that has been suggested by Y.Belyayev (Gnedenko et al., 1969, 1983). Let $\mathrm{P}_k = (0,1) \times (0,1) \times \ldots \times (0,1)$ denote k-dimensional unit cube. It is our parametric space. If $I_i = \{0,1,\ldots,n_i\}$ then $\mathrm{T} = I_1 \times I_2 \times \ldots \times I_k$ is our sample space: $\mathrm{S} = (S_1, S_2, \ldots, S_k) \in \mathrm{T}$. For each sample $\mathrm{S} \in \mathrm{T}$, we determine such a set $C(\mathrm{S}) \subset \mathrm{P}_k$ that

$$P_p\{\mathrm{p} \in C(\mathrm{S})\} \ge \gamma \quad \forall \mathrm{p} \in \mathrm{P}_k . \tag{2.1}$$

For fixed sample $S \in \mathrm{T}$, we find a solution $\underline{\theta}(\mathrm{S})$ of optimization problem

$$\theta_p \to \min \tag{2.2}$$

by constraint $\mathrm{p} \in C(\mathrm{S})$.

Its solution gives the lower γ - confidence limit for θ_p because

$$P_p\{\underline{\theta}(\mathrm{S}) \le \theta_p\} \ge P_p\{\underline{\theta}(\mathrm{S}) \le \theta_p,\ \mathrm{p} \in C(\mathrm{S})\} = P_p\{\ \mathrm{p} \in C(\mathrm{S})\} \ge \gamma \tag{2.3}$$

Now, we need to seek for the criterion function θ_p and system of sets $C(\mathrm{S})$, $\mathrm{S} \in \mathrm{T}$. We begin with the function θ_p.

Let $\Phi(x_1, x_2, \ldots, x_k)$ denote a canonical disjunctive normal form (CDNF) of the boolean function $f(x_1, x_2, \ldots, x_k)$. CDNF Φ contains three Boolean operations: $\vee$ - disjunction, $\wedge$ - conjunction and $\bar{\ }$ - negation and is defined as

$$\Phi(x_1, x_2, \ldots, x_k) = \bigvee_{\alpha \in D(f)} \bigwedge_{i=1}^{k} x_i^{*\alpha_i},$$

where $\alpha = (\alpha_1, \alpha_2, \ldots, \alpha_k) \in B^k$ is k-tuple of zero and unit,

$D(f)$ is unit set of Boolean function f: $\alpha \in D(f) \to f(\alpha) = 1$, $x^{*1} = x$, $x^{*0} = \bar{x}$.

A well-known result (Beichelt & Franken, 1983; Schneeweiss, 1989; Andronov, 1996) says that for mutually independent random boolean variables $X_1, X_2, \ldots, X_k$

$$\theta_p = P_p\{f(X_1, X_2, \ldots, X_k) = 1\} = \sum_{\alpha \in D(f)} \prod_{i=1}^{k} p_i^{*\alpha_i}, \tag{2.4}$$

where $p_i^{*1} = p_i$, $p_i^{*0} = 1 - p_i$.

Also, the value of θ_p can be obtained from CDNF Φ by replacement of disjunction $\vee$, conjunction $\wedge$ and negation $\bar{\ }$ by addition Σ, product $\prod$ and supplement as much as unit $1 - p$, accordingly.

Now, we need to find confidence sets $C(\mathrm{S})$, $\mathrm{S} \in \mathrm{T}$, for θ_p.

The random variables $\{S_i\}$ are mutually independent and S_i has the binomial distribution with known parameter n_i and unknown parameter p_i.

A complementary distribution function of S_i is

$$\overline{F_i}(j) = P\{S_i \geq j\} = \sum_{\nu=j}^{n_i} \binom{n_i}{\nu} p_i^{\nu}(1-p_i)^{n_i-\nu}, \; j = 0,1,\ldots,n_i. \quad (2.5)$$

We introduce the statistic $\overline{F_i}(S_i)$ that is obtained by a substitution the random variable S_i instead of argument j into $\overline{F_i}(j)$. Let us

$$\widetilde{F}(\mathrm{S}) = \overline{F_1}(S_1) \cdot \overline{F_2}(S_2) \ldots \overline{F_k}(S_k). \quad (2.6)$$

Theorem. For every $x \in (0,1)$ holds

$$P\{\widetilde{F}(\mathrm{S}) \geq x\} \leq 1 - \psi_k(x), \quad (2.7)$$

where

$$\psi_k(x) = x \sum_{i=0}^{k-1} \frac{1}{i!}(-\ln x)^i,$$

that can be rewritten as

$$\psi_k(e^{-x}) = e^{-x} \sum_{i=0}^{k-1} \frac{1}{i!} x^i.$$

Proof. The proof is based on the well-known fact (Rao, 1973; Eichenauer-Herrmann, 1995). If a continuous random variable Y has a distribution function $G(y)$ then $G(Y)$ and $1-G(Y)$ have uniform distribution in the interval $(0,1)$. Now, let $R_1, R_2, \ldots, R_k$ be mutually independent random variables with the uniform distribution $(0,1)$. Using a mathematical induction for k, we obtain for every $x \in (0,1)$:

$$P\left\{\prod_{i=1}^{k} R_i < x\right\} = \psi_k(x). \quad (2.8)$$

For every discrete random variable S_i, we introduce a continuous random variable $Y_i = S_i - R_i'$, where $\{R_i'\}$ are mutually independent and uniformly distribution in $(0,1)$. Let $G_i(y)$ denote a distribution function of Y_i. For $j = 0,1,2,\ldots$ and $j-1 < y \leq j$, we have

$$G_i(y) = P\{Y_i < y\} = P\{S_i - R_i' < y\} = P\{S_i < y + R_i'\} =$$

$$= F_i(j) + [F_i(j+1) - F_i(j)](1 - j + y) \geq F_i(j) = F_i(y).$$

Therefore, almost everywhere

$$\overline{F}_i(S_i) = \overline{F}_i(S_i - R_i') = \overline{F}_i(Y_i) \geq \overline{G}_i(Y_i).$$

Using the formula (2.8), we get finally

$$P\{\widetilde{F}(S) \geq x\} = P\left\{\prod_{i=1}^{n} \overline{F}_i(S_i) \geq x\right\} \geq P\left\{\prod_{i=1}^{n} \overline{G}_i(Y_i) \geq x\right\} = 1 - \psi_k(x)$$

,thus, (2.7) holds. (q.e.d.)

Note, that we haven't used the concrete distribution type, therefore, the theorem is true for all integral random variables $\{S_i\}$.

Next, consider the function $\psi(x)$, that is determined by the formula (2.7) for $\gamma \in (0,1)$. It is strictly increasing with a range of a variation $(0,1)$. Let $\psi_k^{-1}(\gamma)$, $\gamma \in (0,1)$, denote its inverse which is strictly increasing, too. Combining (2.1), (2.3) and (2.7) we see that the γ-confidence set $C(S)$ can be determined as

$$C(S) = \{p\colon \widetilde{F}_p(S) \geq \psi_k^{-1}(1-\gamma)\}, \tag{2.9}$$

because

$$P_p\{p \in C(S)\} = P_p\{p \in \{p\colon \widetilde{F}_p(S) \geq \psi_k^{-1}(1-\gamma)\}\} =$$

$$= P_p\{\widetilde{F}_p(S) \geq \psi_k^{-1}(1-\gamma)\} \geq 1 - \psi_k(\psi_k^{-1}(1-\gamma)) = \gamma.$$

Also $\{C(S),\ S \in T\}$ as a system of γ-confidence sets for probability p. Therefore, the components θ_p and $C(S)$ of the optimization problem (2.2) are determined. Here, we study the simplest example (Gnedenko, 1968, 1983) to compare corresponding results.

If a sequential system consist of n elements then its reliability (a probability of no-failure)

$$\theta_p = p_1 p_2 \dots p_k,$$

where p_i is an unknown reliability of the i-th element.

Suppose that the i-th element packed up S_i times in n_i Bernoulli trials. We need to obtain a lower γ-confidence limit $\underline{\theta}$ for probability θ_p.

Mirniy and Solovyev (Gnedenko, 1968, 1983) suggested the lower γ-confidence limit in case when there were no failures $(S_i = n_i \ \forall i)$:

$$\underline{\theta} = \left(1 - \gamma^{1/n_0}\right),$$

where $n_0 = \min\{n_1, n_2, \dots, n_k\}$.

This limit doesn't depend upon the elements number k and is very efficient. For example, if $n_0 = 5$ and $\gamma = 0.5$ then $\underline{\theta} = 0.871$. However, if the failures take place then the lower limit is trivial: $\underline{\theta} = 0$.

In case $\forall S_i = n_i$, our approach gives a problem

$$\theta_p = p_1 p_2 \dots p_k \to \min$$

with constraints $0 \le p_i \le 1 \ \forall i$,

$$p_1^{n_1} p_2^{n_2} \dots p_k^{n_k} \ge \psi_k^{-1}(1-\gamma). \qquad (2.10)$$

This problem has the following solution: if i^* is a number that $n_{i^*} = n_0$ then

$$\underline{\theta} = p_{i^*}^* = \left[1 - \psi_k^{-}(\gamma)\right]^{1/n_0}; \quad p_i = 1, \quad i \ne i^*.$$

For example, if $k = 2$, $\gamma = 0.5$, $n_0 = 5$, then $\psi_5^{-1}(0.5) = 0.1867$ and $\underline{\theta} = (1 - 0.1867)^{1/5} = 0.715$.

If exactly one failure has been fixed in all the trials and it is the i^*-th element, then the constraint (2.10) is replaced by

$$\left(p_{i^*}^{n_0} + n_0 p_{i^*}^{n_0 - 1}\left(1 - p_{i^*}\right)\right) \prod_{i \ne i^*} p_i^{n_i} \ge \psi_k^{-1}(1-\gamma).$$

In this case, $p_i^* = 1$ for $i \ne i^*$, and $\underline{\theta} = p_{i^*}^*$ is a solution of the equation

$$p_{i^*}^{n_0} + n_0 p_{i^*}^{n_0 - 1}\left(1 - p_{i^*}\right) = \psi_k^{-1}(1-\gamma).$$

For the previous data, we get $\underline{\theta} = p_{i^*}^* = 0.5$, $p_i = 1$, $i \neq i^*$. It is essentially better that the trivial limit $\underline{\theta} = 0$.

Generally, determination of overt expression for the limit $\underline{\theta}$ is a very difficult problem. In this connection, we concentrate here on a computational solution of the optimization problem (2.2).

3. Algorithm for a calculation of the lower confidence limit

From the formulas (2.4)-(2.6) and (2.9), we see that the optimization problem (2.2) can be rewritten as

$$\sum_{\alpha \in D(f)} \prod_{i=1}^{k} p_i^{*\alpha_i} \rightarrow \min_{\{p_i\}} \tag{3.1}$$

by constraints $p_i \in (0,1)$, $i = \overline{1,k}$,

$$\prod_{i=1}^{k} \sum_{\nu=S_i}^{n_i} \binom{n_i}{\nu} p_i^{\nu}(1-p_i)^{n_i - \nu} \geq \psi_k^{-1}(1-\gamma). \tag{3.2}$$

For a computational solution of the problem, we use a reduced gradient method (Minox, 1989). It is very efficient for linear constraints. For this, we shall take the logarithm of expressions (3.2) and introduce new variables

$$z_i = z_i(p_i) = -\ln \sum_{\nu=S_i}^{n_i} \binom{n_i}{\nu} p_i^{\nu}(1-p_i)^{n_i-\nu} =$$

$$= -n_i \ln p_i - \ln \sum_{\nu=S_i}^{n_i} \binom{n_i}{\nu}\left(p_i^{-1} - 1\right)^{n_i - \nu}, \quad i = \overline{1,k}. \tag{3.3}$$

Now, the constraint (3.2) is replased by unique linear constraint

$$\sum_{i=1}^{k} z_i \leq -\ln \psi_k^{-1}(1-\gamma). \tag{3.4}$$

The variables p_i belong to the set $(0,1)$, the variables z_i have non-negative values. The function $z_i(p_i)$ increases and has inverse function $p_i = p_i(z_i)$. $p_i(z_i)$ function's values can be calculated by an iterative procedure. If $p_i^{(t)}$ is a value for t -th iterative step, then

$$p_i^{(t+1)} = \exp\left\{-\frac{1}{n_i}\left\{z_i + \ln \sum_{j=S_i}^{n_i} \binom{n_i}{j}\left(\frac{1}{p_i^{(t)}}-1\right)^{n_i-j}\right\}\right\}. \quad (3.5)$$

Since all functions in the formulas (3.1) and (3.2) are continuous and monotone, the minimum in (3.1) is reached at the boundary of the domain (3.2). Therefore the inequality in (3.4) can be replased by an equality and we have the next optimization problem:

$$\varphi(z_1, z_2, \ldots, z_k) = \sum_{\alpha \in D(f)} \prod_{i=1}^{k} (p_i(z_i))^{*\alpha_i} \rightarrow \min \quad (3.6)$$

by constraints $z_i \geq 0$, $i = \overline{1,k}$,

$$g(z_1, z_2, \ldots, z_k) \overset{d}{=} \sum_{i=1}^{k} z_i - b = 0,$$

where $b = \ln \psi_k^{-1}(1-\gamma)$.

To use the reduced gradient method, it is necessary to know derivatives. We have

$$\frac{\partial z_i}{\partial p_i} = -\frac{n_i}{p_i} + \left(\sum_{j=S_i}^{n_i} \binom{n_i}{j}\left(\frac{1}{p_i}-1\right)^{n_i-j}\right)^{-1} \times$$

$$\times n_i \sum_{j=S_i-1}^{n_i-1} \binom{n_i-1}{j}\left(\frac{1}{p_i}-1\right)^{n_i-j-1} \cdot p_i^{-2}. \quad (3.7)$$

The theorem of an inverse function's derivative says that

$$\frac{\partial p_i}{\partial z_i} = \left(\frac{\partial z_i}{\partial p_i}\right)^{-1}\Bigg|_{p_i = p_i(z_i)}. \quad (3.8)$$

The derivative of the criterion function (3.6) respect to z_i is equal to the derivative of the composite function:

$$\frac{\partial \varphi}{\partial z_i} = \frac{\partial \varphi}{\partial p_i}\Bigg|_{p_i = p_i(z_i)} \cdot \frac{\partial p_i}{\partial z_i}. \quad (3.9)$$

To calculate the derivative $\partial\varphi/\partial p_i$, it is comfortable to use the Shannon's decomposition of CDNF Φ respect to the variable x_i:

$$\Phi(x_1,x_2,\ldots,x_k)=x_i \bigvee_{\alpha\in D_i(f)} \bigwedge_{j\neq i} x_j^{*\alpha_j} + \overline{x_i} \bigvee_{\alpha\in \overline{D_i}(f)} \bigwedge_{j\neq i} x_j^{*\alpha_j},$$

where $D_i(f)$ and $\overline{D_i}(f)=D(f)-D_i(f)$ are subsets of unit set $D(f)$ of Boolean function f such as the variable x_i is equal to 1 and 0 correspondingly. Then

$$\frac{\partial\varphi}{\partial p_i}=\sum_{\alpha\in D_i(f)}\prod_{j\neq i} p_j^{*\alpha_j}-\sum_{\alpha\in \overline{D_i}(f)}\prod_{j\neq i} p_j^{*\alpha_j}. \tag{3.10}$$

We introduce some notations:

$$\widetilde{\mathrm{z}}=(z_1,z_2,\ldots,z_{k-1})^T,$$

$$\mathrm{z}=(z_1,z_2,\ldots,z_{k-1},z_k)^T=(\widetilde{\mathrm{z}}^T,z_k)^T,$$

1 - (k-1)-dimensional unit vector,

$$\partial\varphi/\partial\widetilde{\mathrm{z}}=(\partial\varphi/\partial z_1,\partial\varphi/\partial z_2,\ldots,\partial\varphi/\partial z_{k-1}). \tag{3.11}$$

The vector-row

$$\mathrm{u}=(u_1,u_2,\ldots,u_{k-1})=\partial\varphi/\partial\widetilde{\mathrm{z}}-\partial\varphi/\partial z_k\cdot 1$$

is called "a reduced gradient".

We consider the vectors $\widetilde{\mathrm{y}}=(y_1,y_2,\ldots,y_{k-1})^T$ and $\mathrm{y}=(y_1,y_2,\ldots,y_{k-1},y_k)^T=(\widetilde{\mathrm{y}}^T,y_k)^T$, where for $i=\overline{1,k-1}$

$$y_i=\begin{cases}0, & \text{if } u_i>0 \text{ and } \mathrm{z}_i=0,\\ u_i, & \text{otherwise},\end{cases} \tag{3.12}$$

$$y_m=-1\cdot\widetilde{\mathrm{y}}.$$

If it turns out that $\mathrm{y}=0$ the minimum is obtained. Actually, the calculation comes to its end when $\forall y_i<\varepsilon$ where ε is the required accuracy of a solution. If $\mathrm{y} \neq 0$ then it should be found a value $\mu^* > 0$ which minimizes the function $\varphi(\mathrm{z} + \mu\mathrm{y})$ by constraint

$$\mu \leq \mu_{\max} = \min_{y_i < 0}\{-z_i / y_i\}. \tag{3.13}$$

The last problem is solved by an examine of μ values with the increment Δ.

The algorithm of the reduced gradient is given next.

Input: k; (n_i, S_i), $i = \overline{1,k}$; $\Phi(x_1, x_2, \ldots, x_k)$; γ; ε; Δ.

Output: The lower γ-confidence limit $\underline{\theta}$ for θ_p

Step 0. Set $p_i = S_i / n_i$ for $S_i \neq n_i$ and $p_i = 0.9$, otherwise, $i = \overline{1,k}$. Calculate $\{z_i\}$ according to the formula (3.3).

Step 1. Calculate the values of the derivatives by the formulas (3.10), (3.7), (3.8), (3.9).

Step 2. Calculate the reduced gradient u according to the formula (3.11) and form the vector y by the formula (3.12).

Step 3. If $y_i < \varepsilon$ for all $i = \overline{1,k}$ then the algorithm comes to its close; the optimal solution is $(p_1, p_2, \ldots, p_k)$ and $\underline{\theta} = \varphi(z_1(p_1), z_2(p_2), \ldots, z_k(p_k))$. Otherwise, go to the Step 4.

Step 4. Determine $\mu_{\max}$ by the formula (3.13).

Step 5. Minimize the function $\varphi(\mathrm{z} + \mu\mathrm{y})$ by $\mu \in (0, \mu_{\max})$. Let μ^* denote the obtained solution.

Step 6. Form a new value of the vector $\mathrm{z} = \mathrm{z} + \mu^*\mathrm{y}$.

Step 7. Determine a new value of the vector $(p_1, p_2, \ldots, p_k)$, using the iterative procedure (3.5) with $p_i^0 = p_i$. Return to the Step 1.

The presented algorithm has been realized by Master of Science N.Efremova in Pascal. The experience shows so far that the algorithm is efficient enough.

4. Concluding remarks

The suggested algorithm allows numerically to find confidence limits for monotone Boolean functions of random variables. Future investigations are planned to analyze its statistical and computational properties and to extend the present results for a more general class of data.

References

Andronov, A. (1996). Unbiased Estimation of Probability of the Value 1 of a Boolean Function of Random Variables. *Automatic Control & Computer Sciences.* **1**, 21-30.

Beichelt, F., Franken, P. (1983). Zuverlässigkeit und Instandhaltung. *VEB Verlag Technik*, Berlin.

Eichenauer-Herrmann, J. (1995). Pseudorandom Number Generation by Nonlinear Methods. *International Statistical Review*, **63**, 2, 247-255.

Gnedenko, B.V., Belyayev, Y.K., Solovyev, A.D. (1969). Mathematical Methods in Reliability Theory. *Academic Press*, San Diego.

Gnedenko, B.V. et al. (1983). Problems of Mathematical Theory of Reliability. (In Russian). *Radio i Sviaz*, Moscow.

Mann, N.R., Schafer, R.E., Singpurwalla, N.D. (1974). Methods for Statistical Analysis of Reliability and Life Data. *John Wiley*, New York.

Minoux, M. (1983). Programmation Mathematique. Theorie et Algorithmes. *Dunod*, Paris.

Rao, C.R. (1973). Linear Statistical Inferences and its Applications. *John Wiley*, New York.

Schneeweiss, W.G. (1989). Boolean Functions with Engineering Applications and Computer Programs. *Springer-Verlag*, Berlin.

4.4 Concluding remarks

The suggested algorithm allows to construct [illegible] and confidence items of [illegible] Boolean functions of random variables. Future investigations are intended to analyze its statistical and computational properties and to extend the [illegible] measures to a more general type of data.

References

Alexandrov, A. (1994) Unbiased Estimation of Probability of the Value 1 of a Boolean Function of Random Variables. *Automatic Control and Computer Sciences* 4, 21-30

Bauknecht, K., Neumann, P. (1988) Zuverlässigkeit und Instandhaltung. [illegible] *Verlag Technik*, Berlin

Cunningham-Erdmann, J. (1997) Pseudorandom Number Generation in Simulation Methods. *International Statistical Review* 65, 2, 247-255

Gnedenko, B.V., Belyayev, Yu.K., Solovyev, A.D. (1969) Mathematical Methods in Reliability Theory. *Academic Press*, San Diego

Gorodetskii, B.V. et al. (1981) Problems of Mathematical Theory of Reliability (in Russian) *Radio i Svyaz*, Moscow

Mann, N.R., Schafer, R.E., Singpurwalla, N.D. (1974) Methods for Statistical Analysis of Reliability and Life Data. *John Wiley*, New York

Baur, M. (1943) Probabilités [illegible] Théorie et Applications. *Dunod*, Paris

Rao, C.R. (1965) Linear Statistical Inference and its Applications. *John Wiley*, New York

[illegible]

Developing a Graphical Interface for Pre-Posterior Bayesian Analysis

P. ERTO, A. LANZOTTI, M. STAIANO
Dipartimento di Progettazione Aeronautica,
Università di Napoli Federico II
p.le Tecchio 80 - 80125 Napoli - Italy
e-mail: antlanzo@unina.it

Key Words:

Graphical analysis, pre-posterior distribution, weibull reliability estimation

Abstract:

The bayesian approach is a valuable decision support tool for the reliability engineer who faces situations characterised by few and censored experimental data, but usually possesses other knowledge and/or past experience. In fact this statistical procedure, via Bayes theorem, is able to incorporate both professional and statistical pieces of information.

In this work, the authors, developing a prototype of graphical interface, show advantages of graphical analysis of prior and posterior distribution using practical examples of Weibull reliability estimation. Display or plot may help potential user to better control the use of prior information and then understand the results of the bayesian approach, encouraging its use.

INTRODUCTION

The needs of fast and cheap reliability evaluation during the life testing and production of mechanical components encourage the use of Bayesian

*Kitsos, C.P. and Edler, L., Eds.:Industrial Statistics, Physica-Verlag, Heidelberg, 1997

approach. Mainly because this is the only one allowing the reliability engineer to incorporate, into the estimation process, convictions or information that he has before performing the life tests (*see* Mann *et al*, 1974; Martz and Waller, 1982; Erto, 1982).

From a practical point of view, when the data are very few, say 3 or 5, the alternatives to the bayesian estimators are not classical estimators, like the Maximum Likelihood ones, since these very often give estimates that are worse than those suggested by elementary technological considerations. In these cases the practical bayesian approach to decision support works as a filter that improves – in mean – the prior information if these are poor, or substantially confirms them if they are good (Singpurwalla and Song, 1988; Erto and Giorgio, 1994, 1996).

The application of this approach to the Weibull model is meaningful because this is probably the most widely and frequently used life testing model. Further the absence of sufficient statistics for its parameter justifies the large research work spent on this model (Martz and Waller, 1982).

Nevertheless, bayesian approach is not widespread in engineering application. The main reason is confidence loss of not specialised users in results of Bayes theorem application. This is due to practical difficulties both to implement research results or, when implemented, to check the output of computational module. Indeed, for not specialised users, the Bayes theorem is a black "magic" box. To contribute to give light inside the box, this paper shows the results obtained developing a graphical interface for pre-posterior bayesian analysis. The applications emphasise the role of this analysis as a meaningful tool:

1. to match life testing result (or collected "on field" failure data) with the engineering knowledge about the failure mechanism;

2. to evaluate *robustness* of estimation in the face of changing a priori information;
3. to give a practical and user-friendly way to implement bayesian approach and to analyse its results.

In the following paragraphs, the authors present the prototype of graphical interface; after, they give just a brief presentation of bayesian approach to estimation; finally, they show advantages of graphical analysis of prior and posterior distribution using practical examples of Weibull reliability estimation (Erto and Guida, 1985).

GRAPHICAL INTERFACE FOR INPUT AND OUTPUT

The development of the interface for pre-posterior bayesian analysis has the following main aims:

a) to implement and reviewing easily experimental data;
b) to individuate clearly prior information required and to iterate estimation procedure using the same sample data, but changing prior information;
c) to show effectively prior information (as translation of expert opinion) in comparison with *posterior* — obtained using experimental data via Bayes theorem.

Further the interface for input and output has been designed as a pre and postprocessor of bayesian computational modules. In this way, it is adaptable to every kind of formulation of bayesian procedure because of independence from programming language. Information transfer between *I/O* modules and computational one has been realised via exchanging *files*; this way does not reduce global computational efficiency, being just a few data to transfer.

The development of *I/O* modules, based on Windows *Graphical User Interface* (GUI), requires the use of visual development language for the following advantages:

- easy to use for not specialised users too;
- fullness of interface tools (for example, mouse is fully supported);
- growing diffusion in the applications.

Further, visual development language allows overcoming classical limits of the *input* section of computational codes:

- rigid sequentiality in data input;
- hard recognition and correction of mistakes;
- limited reusing of input information for successive data processing.

The development of an input form (Fig. 1) usable via *mouse* enables random data input, while a code verifies coherence showing definition mistakes and allowing data correction. This interactive technique is effective for failure data input, being easy and quick to check and fix mistakes. Input values can be easily modified for successive processing, for example, changing prior information. Further, it is possible to record in a data file input and output information related to previous elaboration for successive use.

Numerical *output* is given trough a synoptic form containing both prior and posterior synthetic information. It is possible to select an option allowing comparison among current results and previous ones.

Further, graphical display of prior distribution enables coherence analysis and comparison with posterior distribution. To this end it has been designed some options — easy to set — to display both prior distribution, that translates expert opinion, and posterior distribution, that is obtained via Bayes theorem application. In this way, it is possible to see the role played by experimental sample of failure data in updating prior knowledge.

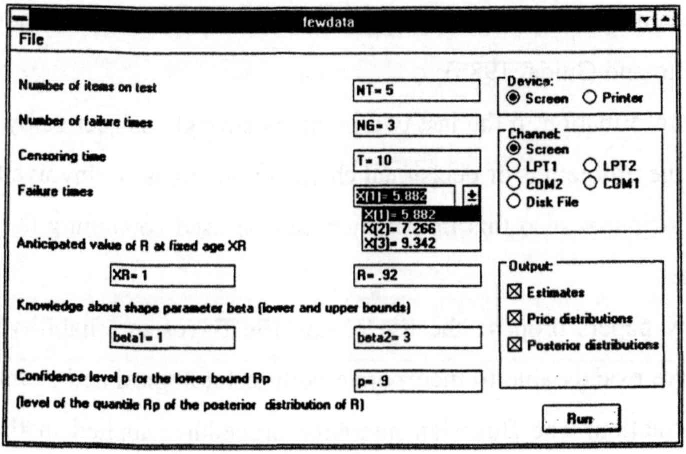

Fig. 1. Input form developed using Windows Graphical Units Interface.

BAYESIAN APPROACH TO WEIBULL RELIABILITY ESTIMATION

Technically speaking the reliability engineer must perform his reliability estimates facing situations characterised by these limited information:

A) few experimental data, since time and costs prevent one to get large statistical samples;

B) "censored" data, since the higher the item reliability is more unlikely the observation time is enough to see its failure (Erto and Lanzotti, 1991).

On the other hand a reliability engineer usually posses other information for the simple reason that (Erto, 1982; Singpurwalla and Song, 1988):

C) in engineering very often a minimum of knowledge exists about the mechanism of failure, under consideration, at least enough to say if the "hazard rate" is increasing, decreasing or constant versus the working time.

D) past experience and/or deterministic calculations usually provide at least the order of magnitude of the reliability R at a fixed age X_R (that a percentage $R \cdot 100$ of the designed items will overcome without failures) (Erto and Guida, 1985).

The information in the last two points is strongly subjective by his nature. But, if the problem is of decisional character and does not involve the consent of a third party, also this information can be used combining them with the first ones.

Many papers propose the use of specific Bayesian reliability estimators, which are exactly able to incorporate both technological and statistical pieces of information. The Bayesian inference procedure, applied in this paper, is presented by Erto and Guida (1985) and has been extended to censored life tests (Lanzotti, 1996).

APPLICATION RESULTS

Let the problem be the estimation of the reliability level R corresponding to a prefixed mission length, x_R, of non repairable parts installed on aircraft. The mission length could be the scheduled renewal time of the part, for example, 100 flight hours.

Redesigning this component to improve its reliability characteristics, experimental data are collected to compare the reliability of new version with old one. A life test of 5 items, censored at 1000 hours, gives 3 ordered lifetimes and 2 survival times:

t_1=588.2	t_2=726.6	t_3=934.2	c_1=1000	c_2=1000

where each t_i (i=1,2,3) is a failure time and each c_j (j=1,2) is a survival time at the end of the life test (I type censored data).

The design engineer evaluates the reliability at 100 hours on the maintenance data (available for the old version) of that component:

$$R(100) = 1 - \frac{\text{number of failures}}{\text{number of working items}} = 1 - \frac{6}{75} = 0.92$$

This estimate can be used as *a priori* information:

$$R_{prior}(100) = 0.92$$

Further, design engineer supposes a wear failure mechanism so he anticipates for the *shape* parameter β of Weibull distribution the prior interval:

$$[1.0,3.0]$$

The posterior expectation of R is:

$$R_{post}(100) = E(R|X) = \int_0^1 R \int_{\beta_1}^{\beta_2} f(R,\beta|X)\, d\beta\, dR = 0.96 \tag{9}$$

and the 90%-quantile, R_{90}, of the posterior distribution of R is given by the equation:

$$\int_{R_{90}}^{1} \int_{\beta_1}^{\beta_2} f(R,\beta|X)\, d\beta\, dR = 0.90 \tag{10}$$

using a numerical procedure to attain the solution of (10), the bound R_{90} at the 90%-confidence level is obtained:

$$R_{90} = 0.93$$

Both the estimates confirm the hypothesis of reliability improvement. To check the robustness of this result, it is very useful to verify *what if* more conservative prior information was used in the estimation procedure. Table 1 shows the results obtained using conservative prior too. These estimates give even more confidence in the previous results.

Figure 2 shows the *posterior* probability density function obtained on the base of these experimental data and different prior information (case 1 and 2), reported in Table 1. Also if the posterior distribution is more negatively skewed in case 2 rather than in case 1, the figure shows *robustness* of posterior in the face of changing prior (*i.e.* expert opinion). The posterior mode appears more robust to different prior information rather than the posterior expectation.

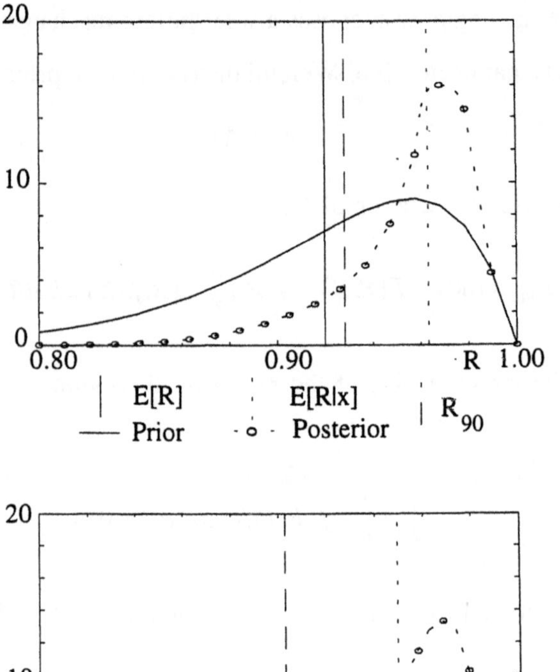

Fig. 2. *Robustness* of *Posterior* distribution of the reliability R at fixed age (X_R) in the face of changing from good prior information (case 1, Tab. 1) to conservative (poor) prior (case 2, Tab. 1)

Tab. 1: Examples of prior and posterior information

	anticipated value for R_{100}	Prior information	Posterior information
case 1	0.92	based on maintenance data	$R_{post}(100) = 0.96$, $R_{90} = 0.93$
case 2	0.80	conservative	$R_{post}(100) = 0.95$, $R_{90} = 0.90$

Graphical testing of prior information

To highlight the behaviour of the posterior distribution for the *shape* parameter β, let us consider pseudo-random sample of n=5 failure times, from a Weibull sampling distribution of parameter R_{100}=0.90 and β=2.5:

t_1=78.7	t_2=185.1	t_3=220.3	t_4=272.1	t_5=347.2

Assuming a good prior information for R_{100} (centred on the sampling value of the parameter), Table 2 shows three cases of prior interval for β:

a. centred and wide interval; the designer knows that the failure mechanism is for wear or fatigue; he has never been estimated β greater than 4; for these reasons, he supposes that β is greater than 1 and less than 4.

b. lower interval, that correspond to poor prior information; the true sampling value is the superior limit of the prior interval.

c. upper interval, that correspond to poor prior information; the true sampling value is the inferior limit of the prior interval.

Tab. 2: Different prior information for the *shape parameter* β

	$[\beta_1, \beta_2]$	Interval type
case *a*	1.0,4.0	centred
case *b*	1.0,2.5	lower
case *c*	2.5,4.0	upper

Figure 3 shows the graphical pre-posterior analysis between the Uniform prior and the resulting posterior distribution in the case *a*. The mode of this distribution is centred in the prior interval and the posterior distribution is symmetric. This is an index of agreement between prior and posterior information.

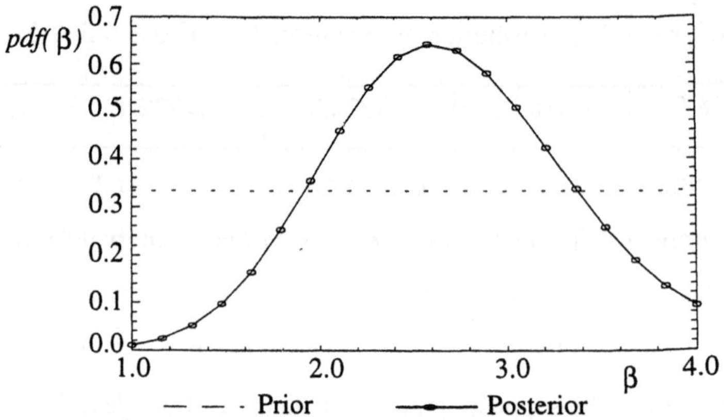

Fig. 3. Prior and Posterior distribution of the shape parameter β for centred (good) prior (case *a*, Tab. 2).

Figure 4, for case *b* and case *c*, shows skewed posterior distribution; the mode is shifted towards superior and inferior limit of the prior interval respectively. In particular, the skewness is more in case *b* than in *c*, it suggests that estimation is more robust in the face of poor under-evaluated prior interval rather than over-evaluated one. Both cases indicate to pay attention to prior information suggesting to extend prior interval because posterior mode is on prior interval limit. Being a pseudo-random sample from a sampling distribution in this case the mismatch between prior and posterior is certainly due to prior. In general, when there is not agreement between prior and posterior, the experimenter has to pay attention to the life testing data too.

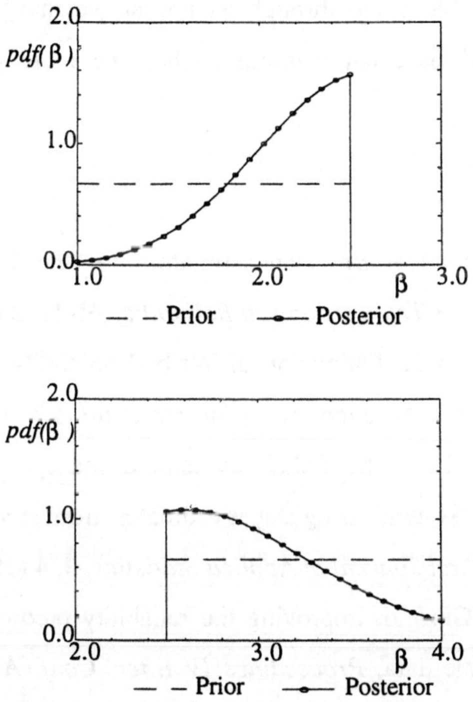

Fig. 4. Prior and Posterior distribution of the shape parameter β for not-centred (good) prior (case *b* and *c*, Tab. 2).

In a certain sense, this graphical analysis plays a role as test of goodness "by eye" of prior information and allows the experimenter both updating his knowledge or improve the experimental life testing condition.

CONCLUDING REMARKS

The whole approach emphasises the role of graphical analysis of *prior* and *posterior* information as a meaningful decision support tool. In fact, the graphical analysis of the results clear shows the improvement in the expert opinion, if any, after the life testing phase. Nowadays, this analysis has no computational problem and through an interactive interface could help the potential users of bayesian methods to become more confident in theirs results.

REFERENCES

Erto, P., New practical Bayesian estimators for the 2–parameter Weibull distribution, *IEEE Transactions on Reliability*, **R-31**, 2 (1982) 194-197.

Erto, P. and M. Guida, Estimation of Weibull reliability from few life tests, *Quality and reliability engineering international*, **1**,3 (1985) 161-164.

Erto, P. and A. Lanzotti: Reliability-centred management of a "Flexible Manufacturing System" using Bayes estimates from small censored samples (in Italian), *Italian Journal of Applied Statistics*, **3**, 4 (1991) 467-475.

Erto, P. and M. Giorgio, Improving the reliability prediction using very few Weibull life time data, *Proceedings IV Inter. Conf. ATA*, Firenze, (1994) 475-487.

Erto, P. and M. Giorgio, Modified 'Practical Bayes-Estimators', *IEEE Transactions on Reliability*, **R-45**, 1 (1996) 132-137.

Lanzotti, A., Graphical Analysis of Weibull Reliability Estimation from few censored life tests, *Int. Conf. on Material Engineering*, Gallipoli-Lecce, (1996) 1097-1104.

Mann, N.R., R.E. Shafer and N.D. Singpurwalla, *Methods for statistical analysis of Reliability and Life data* (Wiley, New York, 1974).

Martz, H.F. and R.A. Waller: *Bayesian Reliability Analysis* (Wiley, New York, 1982).

Shigley, J.E., *Mechanical engineering design* (McGraw-Hill Kogakusha Ltd. 1977).

Singpurwalla, N.D. and M.S.Song: Reliability Analysis Using Lifetime Data and Expert Opinion, *IEEE Transaction on Reliability*, **37**, 3, (1988) 340-347.

Part III

Experimental Design

Nonlinear-Optimal-Sequential Experiment Designs and Applications

CHRISTOS P. KITSOS
Department of Statistics,
Athens University of Economics and Business,
76 Patission Str., Athens 10434 Greece

ABSTRACT

The non-linear experimental design problem is regarded when the errors of observations are assumed iid and normally, distributed as usually. The sequential approach is adopted to reach an optimum experiment design, which is defined through the average per observation iformation matrix. Different applications are discussed and the simulated results presented provide evidence that the sequential approach might be the panacea for solving a non-linear optimal experimental design problem

Key-words: Experimental design, optimal design, Fisher's information, autoregressive model.

1. Introduction

The target of this paper is to refer on the Experiment Design problem when the underlying model describing the physical phenomenon is assumed to be a nonlinear one, in terms of the parameters involved. Different optimality criteria can be introduced, and this is discussed extensively in section 3. These are actually the most commonly used in linear case, Pukelsheim (1993).

The Nonlinear Optimal Experiment Design problem depends heavily on the parameters involved, as the average per observation information matrix depends on the parameters, Ford et.al. (1989), Chaudhuri and Pykland (1995a,b). Therefore the sequential principle of design is adopted to overcome this parameter-dependence. Kitsos (1989), and the method is discussed in sections 4 and 5. In nonlinear situation it is still a problem the optimality of the limiting design, while the development of locally optimal designs has been extensivelly discussed for a certain family of models, Kitsos (1995b), Sitter and Torsney (1995). The interest of this paper is focused on this limiting design which can be D-optimal.

Still the most common applications from biological assays are based on D-optimality, as well as the experimental carcinogenesis, while in Chemistry D-optimality appears an aesthetic appeal, Kitsos (1992, 1995a,b).

Different problems have been discussed under c-optimality: the well known calibration problem, Kitsos (1992) and the rhythmometry problem from the "biological time series", Kitsos et.al. (1988). Both problems can be generalized when a robust approach is adopted, Kitsos and Muller (1995a,b).

2. Background

When an experiment is performed n times, within the experimental region $X \subseteq \Re^k$, known as design space, the outcome or response variable, Y, can be considered either as discrete variable or continuous variable. That means the set of all possible response outcomes Ψ, known as response space, can follow one of the two cases:

Case 1 : Ψ is finite, i.e. $\Psi = \{0,1,2,\ldots,\lambda\}$ with cardinal number $\nu = \lambda+1$. The most common case, $\nu=2$, corresponds to binary response problems where $Y \in \Psi = \{0,1\}$. When n>2 we are refering to polytomous experiments. In some cases, like Poisson experiments, the set Ψ is countable infinite.

Case 2 : Ψ has the power of the continuum, i.e. cardinality c, as Ψ can be any interval in $\Re$.

In the former Case 1 and for binary problems the outcome $Y_i=0$ or 1, $i = 1,2,\ldots n$ is linked with the covariate $x \in X$ and the parameter vector θ, from the parameter space $\Theta \subseteq \Re^p$, through a probability model T as

$$p(x) = p\,(Y_i=1 \mid x) = T(x;\theta) = 1 - P(Y_i=0|x) \qquad (2.1)$$

It is assumed that T is monotonic differentiable function. In bioassay the typical situation is to consider logit, probit or exponential models as T i.e.

$$T_L\,(x,\theta) = \log\,\{p(x)\,(1-p(x))\,\} \quad T_P\,(x,\theta) = \Phi^{-1}\,\{p(x)\} \quad T_E\,(x,\theta) = \exp\,(-\theta x)$$

respectively, with "log", "exp" and "Φ" having their trivial meaning.

Now the latter (Case 2) situation is faced with the general regression model: A real valued, continuous, twice differentiable function, from $X \times \Theta$, is considered to define the (assumed correct) deterministic portion $g(x,\theta)$. The error is applied as a stochastic portion additevly in the form

$$Y_i = g(x_i,\theta) + e_i\,. \quad i = 1,2,\ldots,n\,, \quad \theta \in \Theta \subseteq \Re^p\,, \; x \in X \subseteq \Re^k\,. \qquad (2.2)$$

When a real function f exists such that $g(x_i,\theta)=f(x_i)^T\theta$ then the problem is reduced to the linear regression problem. The non-linear optimal design can be defined through two different problems:

Problem 1 : The underlying model describing the physical phenomenon is nonlinear, as in (2.1), (2.2). The target then is estimate $\theta \in \Theta$ as well as possible.

Problem 2 : A non-linear function, known as general non-linear aspect, of the unknown parameter θ, $\varphi(\theta)$ say, is asked to be estimated as well as possible, even when the underlying model is assumed linear.

In both problems interest is focused on the assumption about the errors. In this paper it is assumed that e_i are iid from the normal distribution $N(0, \sigma^2)$. For a robust approach see Kitsos and Mueller (1995a, b).

In the sequence it would be assumed that the parameter space $\Theta \subseteq \Re^p$ is compact. This assumption is necessary, although not often imposed, when limiting results for the sequence of estimators are obtained.

Unfortunatelly neither Ford, Kitsos and Titterington (1989) nor Chaudhuri and Pykland (1995a) stated that, when the sample size is small, the likelihood might not provide estimates. That occurs when only a batch of successes or failures are in the outcome. Simulation results in Kitsos (1989) for $T_E(x,\theta)$ model provided evidence that, in practice, even with n=50 two or three cases out of one thousand simulations might appear to be pathological. The situation is getting worse with smaller n. Roughly speaking the likelihood estimator exists when, at some design points, the response might be, for some experiments success and for some experiments failure, see Silvapulle (1981). The existence of estimators for models (2.2) have been discussed by Jennrich (1969) in his pioneering work.

Two different lines of assumptions have been imposed for the error variable e. The traditional one is to consider iid errors from the normal distribution with mean zero and variance σ^2, Kitsos (1992, 1995a,b), Ford, Kitsos and Titterington (1989), Chaudhuri and Pykland (1995a,b). Actually, in principle, $\sigma^2=\sigma^2$ $(x; \theta)$ in non-linear situation and $\sigma^2=\sigma^2(x)$ in linear. Robust estimation of a general non-linear aspect $\varphi(\theta)$, as in Problem 2 above based on one-step-M-estimators with a bounded asymptotic bias was discussed by Kitsos and Muller (1995a,b).

The target of this paper is to discuss for the non-linear design problem:

- When a design is optimal i.e. possible optimality criteria that can be imposed.
- How this optimal can be reached i.e. possible computational procedures
- Applications.

The next paragraph discuss the imposed optimality criteria.

3. Optimality Criteria

For both models (2.1) and (2.2) we shall denote by η=E(Y). Then it is

$$\eta = E(Y) = \begin{cases} g(x,\theta) & \text{models (2.2)} \\ T(x,\theta) & \text{models (2.1)} \end{cases} \qquad (3.1)$$

Let $\nabla\eta$ denote the vector of partial derivatives of η with respect to $\theta \in \Theta \subseteq \Re^p$. Then for the exponential family of models Fisher's information matrix is defined to be

$$I(\theta,x) = \sigma^{-2}(\nabla\eta)(\nabla\eta)^T . \qquad (3.2)$$

Moreover if $\eta = \eta(\theta^T x)$ i.e. the covariates and the parameter appear together linearly, then

$$\nabla\eta = [w(\theta^T x)]^{1.2}\, x \qquad \text{with} \quad w(z) = \left[\frac{\partial \eta}{\partial z}\right]^2, \quad z = \theta^T x . \qquad (3.3)$$

Therefore for this particular subcase the information matrix is

$$I(\theta.x) = \sigma^{-2}\, w(\theta^T x) uu^T . \qquad (3.4)$$

Example 1 : Let $P(Y=1) = T(\theta^T x)$, and $\theta_1+\theta_2 x_1 = z = \theta^T x$, with $x=(1,x_1)$. Then the log-likelihood l will be

$$l = \log \{T(z)^Y [1-T(z)]^{1-Y} \} + \text{const}$$

Therefore $I(\theta,x)$ is evaluated as

$$I(\theta.x) = \alpha\,(\theta)uu^T , \quad \text{with} \ \ \alpha\,(\theta) = T'^2[T(1-T)].$$

Notice that T can be either the logit or probit function.

From the above discussion and (3.4) the following is true.

Proposition 1 : For the exponential family of models there exists a vector v such that

$$I(\theta,x)=vv^T \qquad (3.5)$$

The concept of the average-per-observation information matrix plays an important role to our senario for the definition of t optimality criteria for the non-linear experiment design problem. It is defined for ξ, the design measure, Pukelsheim (1993) to be

$$M(\theta,\xi) = \begin{cases} n^{-1} \Sigma\, I(\theta,x_i), \text{ discrete case} \\ \int_{\mathcal{X}} I(\theta,x)\, \xi(dx), \text{ continuous case} \end{cases} \tag{3.6}$$

On the basis of the experiment the average per observation information matrix $M = M(\theta,\xi)$ is obtained which depends, in principle, on θ. Let Mat(s,p) be the set of s×p matrices and let $Q \in Mat(s,p)$ be a matrix of known coefficients. Interest is focused on $Q\theta$. If rank(Q)=p, when s=p, the matrix Q is nonsingular. If s<p we assume rank(Q)=s. We define the operator J_Q applied to M, with Q as above:

$$J_Q(M) = QM^{-}(\theta,\xi)Q^T \tag{3.7}$$

with M^- a generalized inverse of M and $Q \in Mat(p,s)$.

Let NMat(s,s) be the set of nonnegative definite s×s matrices and ω a real valued, convex, decreasing function, on NMat(s,s). Then the following definition is given:

Definition 1 : The design measure ξ^*, from a family of design measures Ξ, is called ω-optimal if and only if

$$\omega\, \{J_Q\, [M(\theta,\xi^*)\} = \min\{\omega\{QM^{-}(\theta,\xi)Q^T\},\ \xi \in \Xi\} \tag{3.8}$$

Under this definition the φ criterion introduced by Titterington (1980) is

$$\varphi = \omega \circ J_I \tag{3.9}$$

where $\circ$ denotes the composition of two functions, and I denote the identity matrix.

Example 2 : Traditional definitions for ω are

$$\text{logdet}(QM^{-}Q^T)\,,\ \ \text{tr}(QM^{-}Q^T)\,,\ \ \text{max eigenvalue}\,(QM^{-}Q^T),$$

$$\text{suptr}\{I(0,x)QM^{-}Q^T\}$$

and corresponds to $D(\theta)$, $A(\theta)$, $E(\theta)$, $G(\theta)$-optimality.

Typical examples for Q are the following three:

$Q = I = \text{diag}(1,1,\ldots,1) \in Mat(s,s)$

$Q = [\,I_s : 0\,]\,,\ I_s = \text{diag}(1,\ldots,1) \in Mat(s,s),\ 0 = \text{diag}(0,\ldots,0) \in Mat(s, p\text{-}s)$

$Q = c \in Mat(p,1)$

and corresponds to interest on a subset of parameters (e.g. $D_s(\theta)$, $G_s(\theta)$, etc.) and to c-optimality, under the assumption that $M^{-1}(\theta,\xi)$ exists. See Pukelsheim (1993, Chapter 6) for a compact reference for linear case.

Example 3 : Recall Example 1. The D(θ)-optimal design concentrated at two points, i.e. $\xi_1=\xi_2=1/2$ and the optimal design points are

$$x_1 = (x_0-\theta_1)/\theta_2 , \quad x_2 = (-x_0-\theta_1)/\theta_2 .$$

The D-optimal points corresponds to $x_0=1.54$ for the logistic and $x_0=1.14$ for the probit. If the design space $X = [\alpha,\beta]\subseteq\Re$ is symmetric about $-\theta_1/\theta_2$ and $(\pm x_0-\theta_1)/\theta_2 \notin X$ then $x_1=\alpha$, $x_2=\beta$.

Among the optimality criteria D-optimality appears an aesthetic appeal for Chemical Kinetics, Kitsos (1995), while only few problems have been tackled by c-optimality, Kitsos (1989), Kitsos et al. (1988) as the geometry involved is rather repulsive to the experimenter.

The nonlinear theoretical framework is based on the true θ and is such case the so-called equivalence theorems from the linear situation are extended to the nonlinear case. But is practice the situation is different. To get the minimum is rather a difficult computational problem. An approach which seems the most applicable is the sequential approach which is discussed in the next paragraph.

4. Sequential Approach

When the sequential principle of designing is adopting the experimenter selects n points and evaluates an estimate of θ, $\hat{\theta}_n$, say. Then the experimenter redesigns using the vector $\hat{\theta}_n$ and a new estimator is obtained. Three questions are arisen:

- How the initial design is chosen?

- How the sequence of points is created?
- Does the sequence of estimators converge to θ?

We briefly discuss these questions:

The initial design is build up on the locally optimal experiment design, Ford, Kitsos and Titterington (1989). The problem is that the optimum design points in nonlinear problems depends on the parameter under investigation. Therefore a prior knowledge of the neighbourhood of θ is asked. In practice the experimenter can provide such information. The sequence of points $x(n+1) = (x(n),x_{n+1}) = (x_1,x_2,\ldots,x_n,x_{n+1})$ is created by choosing that point, as a next stage, which minimizes the determinant of the covariance matrix for a current estimate of θ. Recall that $nM(\theta,\xi)$ is asymptotically the covariance matrix. Moreover, at each point it is allocated an equal batch of points.

Definition 2 : A sequential procedures which allocates $n_0 = \dim\theta$ observations per stage is called a fully sequential procedure.

Then, the following is true due to Kitsos (1989).

Proposition 2 : If the initial design is D-optimal the limiting design is then a D-optimal, if a fully sequential procedure, through stochastic approximation, is adopted.

The sequence of estimators will converge in mean square to θ, if the stochastic approximation scheme is adopted for models (2.1), (2.2), see Kitsos (1992, 1989) respectively. Now the following can be proved.

Proposition 3 : For models (2.2) the iterative scheme

$$\theta_{n+1} = \theta_n - (\lambda s^2 n^{-1}) \sum_{i=1}^{p} \frac{\mu_i}{\lambda_i} V_i \tag{4.1}$$

with: $\lambda \in (0,1)$, s^2 a suitable estimator of σ^2, $\theta_n \in \Theta \subseteq \Re^2$, λ_i, V_i, $i = 1,2,\ldots,p$ the eigenvalues and the eigenvectors of $M=M(\theta,\xi)$ respectively and μ_j the eigenvalues of ∇g, converges to θ.

Indeed:

Apply Newton and Kantorovitch Theorem, Ortega and Pheinfold (1970) the iterative scheme

$$\theta_{n+1} = \theta_n - (\lambda s^2 (M(\theta,\xi))^{-1} \frac{\partial g}{\partial \theta} \tag{4.2}$$

converges to θ, as $\sigma^2 nM(\theta,\xi)$ is the Hessian, from a Numerical Analysis point of view. Let the eigenvalue decomposition of M, which is in accordance with Proposition 1 and the eigenvector decomposition to be

$$M = \sum \lambda_i V_i V_i^T , \qquad \frac{\partial g(x(n);\theta)}{\partial \theta} = \sum \mu_j V_j . \tag{4.3}$$

As $V_i V_j^T = \delta_{ij}$ then (4.2) is reduced to (4.1).

In the following paragraph we discuss some aplication of the nonlinear optimal design problem through sequential approach.

5. Applications

As the nonlinear experimental design problem suffers "on θ dependence", i.e. $M=M(\theta,\xi)$, there is not a unique theoretical framework for every model (2.1) or (2.2) while Chaudhuri and Pykland (1995a,b) provided limited results for the sequence of the average per observation information matrices. It is rather difficult to consider the sequence $\omega(M(\hat{\theta}_n, \xi_n)$ and see when is converging to $\omega(M(\theta,\xi^*))$.

Most of the application are refering to D-optimality. Examples 4 and 6 and 5 are refering to models (2.1) and (2.2) respectively.

Example 4 : For the carcinogenic experiments, it is consider the model $T_H(x,\theta) =$ 1-exp$\{(\theta_0+\theta_1 x)\}$, known as One Hit model. For this model the 100p percentile, L_p, can be evaluated sequentially, through the iteration

$$L_{p.n-1} = L_{p.n} - (n\hat{\theta}_1 (1-p))^{-1}(y_n - p) \tag{5.1}$$

see Kitsos (1992, 1995a). That is the sequence $L_{p,n+1}$ converges in mean square to L_p.

Example 5 : In chemical kinetics different models have been developed to describe a chemical process. To develop the sequential design the initial local optimum design is needed. for the particular model. Therefore, Kitsos (1995b) provides all the appropriate support points. For the family of Generalized Linear Models the support points have been provided by Sitter and Torsney (1995).

Example 6 : For the dillusion series assessment it is considered that $T_E(x,\theta)=\exp(-\theta x)$, $\theta>0$, $x>0$. It is discussed in Kitsos (1989) that fully sequential procedures obtain better estimated mean square errors (EMSE). In the following Table 1 results from 1000 simulations is appeared comparing EMSE with different sample sizes $n_1=50$, $n_2=75$, when the initial batch size is r=5,25. Moreover, r observations are devoted at the optimum point $x_0=1.59/\theta$ at each stage. Therefore the procedure is sequential but not fully sequential. Different starting values θ_{st} are chosen to feed the procedure. The corresponding estimates $\hat{\theta}_1$ and $\hat{\theta}_2$ are also provided.

Table 1 provides evidence that as the sample size increases better estimates in MSE are obtained. Notice also that when the batch size is small (r=5) the estimators are better as more stages are included in the sequential procedure.

Table 1: Dilution series assessment. Comparison studies of EMSE for sample size n_1=50, n_2=75. True value of θ=3.18.

θ_{st}	r	$\hat{\theta}_1$	$\hat{\theta}_2$	EMSE1	EMSE2
2.196	5	3.31	3.27	.40	.25
	25	3.28	3.24	.41	.24
	50	–	3.33	–	.46
3.18	5	3.26	3.24	.37	.22
	25	3.25	3.22	.36	.21
	50	–	3.25	–	.37
7.15	5	3.28	3.24	.37	.23
	25	3.28	3.25	.41	.25
	50	–	3.22	–	.23

Example 7 : The autoregressive model, discussed also by Ford et al. (1985),

$$y_{i-1} = \theta\, y_i + \varepsilon_{i+1}\,, \quad i = 1,2,\ldots,n$$

appears the features of a fully sequential design, with the "next" design point being y_i. Under the usual normal assumption for the errors the estimate of θ and the sample information is respectively

$$\hat{\theta}_{n+1} = \sum_{i=1}^{n} y_i y_{i+1} \Big/ \sum_{i=1}^{n} y_i^2, \qquad \frac{1}{\sigma^2} I_n = \frac{1}{\sigma^2}\sum_{i=1}^{n} y_i^2$$

It was pointed out by Lai and Siegmund (1983) the following limiting (in Law) result

$$I^{1/2}\left(\hat{\theta}_n - \theta\right) \to N(0,\sigma^2), \quad \text{with } |\theta| < 1 .$$

This result can be put into context by noting that if the design points were fixed in the model $y_i = \theta x_i + e_i$, i=1,2,…,n and $\hat{\theta}_n = \sum y_i x_i / \sum x_i$ then the above limiting result holds exactly for any n and θ. In Table 2 the results of 1.000 simulations with "small" sample sizes, n=10,5 observations, $\theta \in [-1,1]$ are presented and $\bar{\theta}$ being the average estimate of $\hat{\theta}_i$, i = 1(1)1000 and EMSE as in Table 1. Approximate confidence intervals were evaluated and the estimated coverage (EC) for a nominal level $\alpha = 0.05$ are calculated.

As n is getting larger (but still small) $\bar{\theta}$ is getting better, while the value of θ does not influence the EMSE. In both cases the EC is rather in accordance with the nominal level despite the small size.

Table 2: The autoregressive model. Initial value y_1=0.0, σ=1, α=0.05. Comparison on EMSE and EC for sample sizes n_1=10, n_2=5.

θ	$n_1 = 10$			$n_2 = 5$		
	EMSE	$\bar{\theta}$	EC	EMSE	$\bar{\theta}$	EC
−1.0	0.09	−0.84	.947	0.27	−0.82	.969
−0.5	0.09	−0.42	.970	0.25	−0.39	.979
0.0	0.10	0.00	.967	0.25	0.02	.970
0.5	0.10	0.41	.965	0.33	0.79	.973
1.0	0.09	0.85	.955	0.69	0.38	.975

6. Discussion

The nonlinear design problem was considered and the sequential approach is adopting to solve the problem. There is not a global theoretical framework, for every optimality criterion. Most of the theoretical results are refered to D-optimality, keeping the sequence $M(\theta_n,\xi)$ bounded, with no interest on $M(\theta,\xi_n)$ sequence and no results for the sequence $M(\theta_n,\xi_m)$. Still the general problem: Does the sequence $\omega(M(\theta_n,\xi_m))$ converges to $\omega(M(\theta,\xi^*))$, has not been tackled.

References

1. **Chaudhuri, P. and Pykland, P.A. (1995a).** Nonlinear Experiments: Optimal Design and Inference Based on Likelihood. *JASA*, 88, 538-546.
2. **Chaudhuri, P. and Pykland, P.A. (1995b).** On Efficient Designing of Nonlinear Experiments. *Statistica Sinica*, 5, 421-440.
3. **Ford, I., Titterington, D.M. and Wu, C.F.J. (1985).** Influence and sequential design. Biometrika, 72, 545-551.
4. **Ford, I., Kitsos, C.P. and Titterington, D.M. (1989).** Recent advances in nonlinear experiment design. *Technometrics*, 13, 49-60.

5. **Jennrich, R.J. (1969).** Asymptotic properties of non-linear Least Squares estimators. *Ann. Math. Stat.*, 40, 633-643.

6. **Kitsos, C.P., Titterington, D.M. and Torsney, B. (1988).** An optimal design problem in rhythmometry. *Biometrics*, 44, 657-671.

7. **Kitsos, C.P. (1989).** Fully-sequential procedures in nonlinear Design problems. *Comp. Stat. and Data Analysis*, 8, 13-19.

8. **Kitsos, C.P. (1992).** Quasi-sequential procedures for the calibration problem. *In COMPSTAT 1992, Y. Dodge and J. Whittaker (eds.)*. Physica-Verlag, 2, 227-231.

9. **Kitsos, C.P. (1992).** Adopting Sequential Procedures for Biological Experiments. In *MODA3 (Model Oriented Data Analysis), Muller, Wynn, Zhigliavsky (Eds.)*, pg3-9. Physica-Verlag.

10. **Kitsos, C.P. (1995a).** Sequential Assays for Experimental Carcinogenesis. *In ISI 1995, 21-29 Aug., Beijing, China, 50th Session, Two Books(1), 21-29 Aug., Beijing, China.*

11. **Kitsos, C.P. (1995b).** On the support Points of D-optimal Nonlinear Experiment Designs for Chemical Kinetics. In *MODA4 (Model Oriented Data Analysis), C. Kitsos, W. Muller (Eds.)*, pg 71-76.

12. **Kitsos, C.P. Muller, Ch. H (1995a).** Robust estimation of Non-linear Aspects. In *MODA4 (Model Oriented Data Analysis),C. Kitsos, W. Muller (Eds.)*, pg 71-76.

13. **Kitsos, C.P. and Muller, Ch.H. (1995b).** Robust Linear Calibration. *Statistics*, 27, 93-103.

14. **McCallagh, P., Nelder, J. (1989).** *Generalized Linear Models.* Chapman and Hall.

15. **Lai, T.L. and Siegmund, D. (1983).** Fixed accuracy estimation of an autoregressive parameter. *Ann. Stat.*, 11, 478-485.

16. **Muller, Ch.H. (1994b).** On the calculation of MSE minimizing robust estimators. In *COMPSTAT, 1994, R. Dutter, W. Grossman (Eds) pg 257-262.* Physica-Verlag.

17. **Pukelsheim, F. (1993).** *Optimal Design of Experiments.* John Wiley.

18. **Silvapulle, M.J. (1981).** On the Existence of the Maximum Likelihood Estimators for the Binomial Response Models. *JRSS, B*, 43, 310-313.

19. **Sitter, R.R., Torsney, B. (1995).** D-optimal Designs for Generalized Linear Models. In MODA4 *(Model Oriented Data Analysis),C. Kitsos, W. Muller (Eds.)*, pg. 87-102.

20. **Titterington, D.M. (1980).** Aspects of Optimal Design in Dynamic Systems. *Techrometrics*, 22, 287-299.

21. **Wu, C.F.J. (1985).** Efficient sequential Designs with Binary Data. *JASA*, 80, 974-984.

Robust Inference and Experimental Design for Multi-Factor Models

CHRISTINE H. MÜLLER

Freie Universität Berlin
Fachbereich Mathematik und Informatik, WE 1
Arnimallee 2-6, D - 14195 Berlin, Germany
e-mail: mueller@math.fu-berlin.de

Key Words: Multi-factor models, robust estimation, testing, breakdown points, designs.

Abstract: For multi-factor linear models outlier robust estimators and tests are derived. As robustness criteria the finite sample breakdown point and the asymptotic bias in shrinking contamination neighbourhoods are used. It is shown that one-step M-estimators can combine high breakdown point, small asymptotic bias and high efficiency. For tests similar results can be shown by basing the tests on one-step M-estimators. It turns out that highest efficiency can be achieved if appropriate designs are used.

1 Introduction

In many industrial applications the observations of an experiment can be influenced by a large number of factors. Then the aim of the statistical analysis is to identify the factors which have an influence, to quantify the influence and to find those levels of the factors which provide maximum output.

In many situations the relationship between the observations and the experimental conditions, given by the different levels of the factors, can be described by a linear model. In a general linear model the vector $Y = (Y_1, ..., Y_N)'$ of all observations is given by

$$Y = X(d)\beta + Z, \tag{1.1}$$

*Kitsos, C.P. and Edler, L., Eds.:Industrial Statistics, Physica-Verlag, Heidelberg, 1997

where $Z = (Z_1, ..., Z_N)'$ is the vector of error variables, $\beta \in I\!R^r$ an unknown parameter vector, $d = (t_1, ..., t_N)' \in \mathcal{T}^N$ the vector of known experimental conditions, i.e. the design, $X(d) = (x(t_1), ..., x(t_N))' \in I\!R^{N \times r}$ the design matrix, and $x : \mathcal{T} \to I\!R^r$ the vector of known "regression" functions. Realizations of the N-dimensional random vectors Y and Z are denoted by $y = (y_1, ..., y_N)'$ and $z = (z_1, ..., z_N)'$, respectively.

In multi-factor linear models an experimental condition t has the form $t = (\tau_1, \ldots, \tau_K) \in \bigotimes_{k=1}^K \mathcal{T}_k$, where $\mathcal{T}_k \subset I\!R$ is the set of all possible levels of the factor τ_k. $\mathcal{T}_k \subset I\!R$ can be finite but also an interval. $x(t)$ can also be expressed by the different factors. Let $x_1 \otimes x_2$ denote the Kronecker product of x_1 and x_2. Then in a model with all interactions we have

$$x(t) = x_1(\tau_1) \otimes x_2(\tau_2) \otimes \ldots \otimes x_K(\tau_K). \tag{1.2}$$

For example in (1.2) we can have $x_k(\tau_k) = (1, \tau_k, \tau_k^2, \ldots, \tau_k^{r_k})$ so that $r = \Pi_{k=1}^K (r_k + 1)$ is the dimension of the model. In a model with no interactions the regressors have the form

$$x(t) = (1, x_1(\tau_1)', x_2(\tau_2)', \ldots\ldots, x_K(\tau_K)')', \tag{1.3}$$

where for example $x_k(\tau_k) = (\tau_k, \tau_k^2, \ldots, \tau_k^{r_k})$ so that $r = (1 + \sum_{k=1}^K r_k)$ is the dimension of the model. Models with partial interactions are also possible.

To quantify the influence of the factors means to estimate the whole vector β or to estimate components of β. In general, components of interest can be expressed by $L\beta$, where $L \in I\!R^{s \times r}$ is given. Another aim of statistical analysis is to test special hypotheses on the components of β. For example to test the hypothesis that special interactions are equal to zero. In general such hypotheses can be expressed by $H_0 : L\beta = l$, where $L \in I\!R^{s \times r}$ and $l \in I\!R^s$ are given.

If the errors $Z_1, \ldots, Z_N$ are independent and identically distributed according to a distribution P with mean 0 and variance σ^2, then the Gauss-Markov estimator is the best linear unbaised estimator. Moreover, if P is the normal distribution $n_{(0,\sigma^2)}$, then the F-tests is the best invariant test. But if outliers appear, then the Gauss-Markov estimator and the F-test can be biased very much so that alternative estimators and tests should be used.

In Section 2 we regard the question of outlier robust and efficient estimation of $L\beta$ in multi-factor linear models. In Section 3 we discuss the problem of outlier robust and efficient testing of $H_0 : L\beta = l$. For estimation as well as for testing we will see that an efficient robust statistical inference is supported by appropriate designs.

2 Robust and efficient estimation

The Gauss-Markov estimator $\widehat{\varphi}_{LS}$ for $\varphi(\beta) = L\beta$ has the form $\widehat{\varphi}_{LS} = L\widehat{\beta}_{LS}$, where $\widehat{\beta}_{LS}$ is the least squares estimator for β, i.e.

$$\widehat{\beta}_{LS}(y,d) = \arg\min_{\beta} \sum\nolimits_{n=1}^{N} (y_n - x(t_n)'\beta)^2.$$

Because outliers produce large residuals in robust estimators the influence if large residuals should be reduced. This can be done by cutting residuals which exceed special values or by dropping a given number of largest residuals. Dropping the $N-h$ largest residuals leads to the h trimmed L_p estimators $\widehat{\varphi}_h = L\widehat{\beta}_h$ with

$$\widehat{\beta}_h(y,d) = \arg\min_{\beta} \sum\nolimits_{n=1}^{h} w_n R_{(n)}(y,d,\beta)^p,$$

where $p > 1$, $w_1, \ldots, w_h > 0$ and $R(y,d,\beta) = (R_1(y,d,\beta), \ldots, R_N(y,d,\beta))'$ with $R_n(y,d,\beta) = |y_n - x(t_n)'\beta|$. $R_{(1)}(y,d,\beta) \leq R_{(2)}(y,d,\beta) \leq \ldots \leq R_{(N)}(y,d,\beta)$ denote the ordered residuals. Compare with Coakley and Mili (1993), Mili and Coakley (1993), Müller (1995a,b, 1997). Cutting the residuals with a score function $\rho : I\!R \times \mathcal{T} \to I\!R$ leads to the M-estimators $\widehat{\varphi}_\rho = L\widehat{\beta}_\rho$, where

$$\widehat{\beta}_\rho(y,d) = \arg\min_{\beta} \sum\nolimits_{n=1}^{N} \rho(y_n - x(t_n)'\beta, t_n) \tag{2.4}$$

(see e.g. Huber 1981, Hampel et al. 1986). M-estimators also can be used in a one-step version. Such estimators $\widehat{\varphi}_\psi$ are given by (see Müller 1994b)

$$\widehat{\varphi}_\psi(y,d) = L\widehat{\beta}(y,d) + \tfrac{1}{N} \sum\nolimits_{n=1}^{N} \psi\left(\tfrac{y_n - x(t_n)'\widehat{\beta}(y,d)}{\widehat{\sigma}(y,d)}, t_n\right) \widehat{\sigma}(y,d),$$

where $\widehat{\beta}$ is an initial estimator for β and $\widehat{\sigma}$ is an estimator for the standard deviation σ. $\psi : I\!R \times \mathcal{T} \to I\!R$ is the derivative of the original score function ρ. The advantage of one-step M-estimators is that different robustness and efficiency properties can be included by using the initial estimator $\widehat{\beta}$ and the score function ψ. In particular with the initial estimator $\widehat{\beta}$ we can achieve high breakdown points at finite samples, and with the score function ψ we can achieve high asymptotic robustness and high asymptotic efficiency.

Breakdown points. The breakdown point of an estimator $\widehat{\varphi}$ is the smallest proportion of outliers which can cause an arbitrary large bias of the estimator, i.e.

$$\epsilon^*(\widehat{\varphi}, y, d) = \frac{1}{N} \min\{M;\ \sup_{\tilde{y} \in \mathcal{Y}_M} |\widehat{\varphi}(y, d) - \widehat{\varphi}(\tilde{y}, d)| = \infty\},$$

where

$$\mathcal{Y}_M = \{\tilde{y} \in I\!R^N;\ \ \text{card}\{n;\ \tilde{y}_n \neq y_n\} \leq M\}$$

(see He et al. 1990, Ellis and Morgenthaler 1992). Jurečková and Portnoy (1987) showed that in the case $L = E_{r \times r}$ ($E_{r \times r}$ denotes the identity matrix) the breakdown point of the initial estimator $\widehat{\beta}$ is transferred to the one-step M-estimator $\widehat{\varphi}_\psi$, if $\widehat{\sigma}$ has at least the same breakdown point and ψ is bounded. For general $L \in I\!R^{s \times r}$ it can be easily seen that the breakdown point of the one-step M-estimator can be only improved, because always we have $\epsilon^*(L\widehat{\beta}, y, d) \geq \epsilon^*(\widehat{\beta}, y, d)$. Usually it is easy to find an estimator $\widehat{\sigma}$ with a breakdown point not much smaller than 1/4 (see Rousseeuw and Croux 1993). Since in high dimensional models, as it is the case for the multi-factor models, the highest possible breakdown points are often approximately $\frac{1}{r} << \frac{1}{4}$, the main problem is to find initial estimators $\widehat{\beta}$ with high breakdown points.

In Müller (1995a) it was shown that the highest breakdown points are achieved by h-trimmed L_p estimators $\widehat{\beta}_h$ with $[\frac{N+\mathcal{N}(d)+1}{2}] \leq h \leq [\frac{N+\mathcal{N}(d)+2}{2}]$, where $[z] = \max\{n \in I\!N;\ n \leq z\}$ and $\mathcal{N}(d) = \sup_{\beta \neq 0} \text{card}\{n;\ x(t_n)'\beta = 0\}$ is the maximum number of regressors in a subspace. In Mizera and Müller (1996) it was shown that the highest breakdown points are also attained by some M-estimators. Such an M-estimator is for example the Cauchy estimator, which has the score function $\rho(z, t) = \log(1 + z^2)$.

For calculating h-trimmed L_2 estimators there exist very accurate algorithms (see Müller 1996). One is also implemented in the statistics package S-PLUS under the name "ltsreg". The only problem is then to determine $\mathcal{N}(d)$ for the ideal trimming proportion, which in particular can be in models with many factors a problem. To avoid this problem we can use the Cauchy estimator which automatically gives the maximum breakdown point and which is also implemented in S-PLUS. But for this implementation it is not clear if it really gives the minimum as requested by (2.4) or if it provides a locally extremum.

Asymptotic robustness and efficiency. With the score function ψ of an one-step M-estimator the asymptotic robustness and efficiency properties can be determined. For that we assume that the design d is converging to an asymptotic design measure in the following sense: $\lim_{N\to\infty} \frac{1}{N}\sum_{n=1}^{N} 1_t(t_n) = \delta(\{t\})$ for all $t \in \mathcal{T}$. To model the asymptotic influence of outliers we assume that the error Z_n/σ has a distribution of the form

$$(1 - RN^{-1/2}\epsilon(t_n))n_{(0,1)}(dz) + RN^{-1/2}\epsilon(t_n)g(z,t_n)n_{(0,1)}(dz), \qquad (2.5)$$

i.e. the ideal normal distribution $n_{(0,1)}$ is contaminated by $g(\cdot,t_n)n_{(0,1)}$, where $g(\cdot,t_n)$ is an arbitrary density. The proportion of contamination is given by $R > 0$ and $\epsilon : \mathcal{T} \to [0,1]$. The maximum asymptotic bias caused by such contamination is for a one-step M-estimator

$$R\,\|\psi\|_\delta, \quad \text{where } \|\psi\|_\delta = \text{ess} \sup_{n_{(0,1)}\otimes\delta} |\psi(z,t)|.$$

The asymptotic covariance matrix of the one-step M-estimator is under the ideal model as well as under the contaminated model

$$C(\psi,\delta) = \int \psi(z,t)\psi(z,t)'n_{(0,1)}(dz)\delta(dt).$$

See Rieder (1985, 1987, 1994), Kurotschka and Müller (1992), Müller (1995b, 1997). To obtain robust and efficient estimators one can minimize the trace of the asymptotic covariance matrix $C(\psi,\delta)$ under the robustness constraint that $\|\psi\|_\delta \leq b$. For general designs the solution can be given only very implicitly (see Kurotschka and Müller 1992). But for A-optimal designs and for designs with minimum support the optimal score function ψ_b has the form

$$\psi_b(z,t) = L\,I(\delta)^- x(t)\,\mathrm{sign}(z)\frac{\min(|z|, b\,v_b(t))}{|L\,I(\delta)^- x(t)|\,v_b(t)}, \tag{2.6}$$

where $I(\delta) = \int x(t)x(t)'\delta(dt)$, $v_b(t) = \frac{2\Phi(b\,v_b(t))-1}{|L\,I(\delta)^- x(t)|} > 0$ and Φ denotes the distribution function of the standard normal distribution. The fixed points $v_b(t)$ can be easily calculated by Newton's method. If δ is an A-optimal design for $L\beta$, i.e. $\delta \in \arg\min_{\tilde{\delta}} \mathrm{tr}(L\,I(\tilde{\delta})^- L')$, then only one fixed point v_b has to be determined. Moreover, at A-optimal designs the maximum efficiency is achieved. See Müller (1994a, 1995b, 1997).

In general multi-factor models A-optimal designs are difficult to calculate. But in models with all interactions, no interactions or with complete M-factor interactions A-optimal designs can be calculated as products of the A-optimal designs for the single factor models (see Schwabe, 1996). But also A-optimal designs can have a disadvantage. Namely, for designs d approaching an A-optimal design it is sometimes difficult to determine $\mathcal{N}(d)$ for the optimal trimming proportion of the initial trimmed L_2-estimator. Then the Cauchy estimator should be used as initial estimator. This problem is avoided by designs with minimum support because for such designs it is easy to determine $\mathcal{N}(d)$.

Example: Quadratic regression. Assume $x(t) = x(\tau_1,\tau_2) = x_1(\tau_1)\otimes x_2(\tau_2) = (1,\tau_1,\tau_1^2,\tau_2,\tau_1\tau_2,\tau_1^2\tau_2,\tau_2^2\ \tau_1\tau_2^2,\tau_1^2\tau_2^2)'$, where $x_i(\tau_i) = (1,\tau_i,\tau_i^2)'$ for $i = 1,2$. Assume also $\beta = (\beta_0,\beta_1,\ ..,\beta_8)'$ and that $L\beta = (\beta_0,\beta_1,\beta_3,\beta_4)'$, i.e. only the lower order terms without the quadratic effects are of interest. Let $\beta^i = (\beta_0^i,\beta_1^i,\beta_2^i)'$ denote the unknown parameters in the marginal models given by x_i. Then for $L^i = \begin{pmatrix} 1 & 0 & 0 \\ 0 & 1 & 0 \end{pmatrix}$ we have $L^i\beta^i = (\beta_0^i,\beta_1^i)'$ for $i = 1,2$ and $L = L^1\otimes L^2$. It is known that $\delta^i = \frac{1}{4}e_{-1} + \frac{1}{2}e_0 + \frac{1}{4}e_1$ is the A-optimal design for $L^i\beta^i$ in $\Delta^i = \{\delta;\ \mathrm{supp}(\delta)\subset[-1,1]\}$. Hence, according to Corollary 4.6 in Schwabe (1996) $\delta = \delta^1\otimes\delta^2$, i.e. the design which assigns weights $\frac{1}{16}$ to the four corner points $(-1,-1)$, $(-1,1)$, $(1,-1)$ and $(1,1)$, weights $\frac{1}{8}$ to the four mid-points of the edges $(-1,0)$, $(0,-1)$, $(1,0)$ and $(0,1)$ and weight $\frac{1}{4}$ to the center point $(0,0)$, is A-optimal for $L\beta$ within $\Delta = \{\delta;\ \mathrm{supp}(\delta)\subset[-1,1]\times[-1,1]\}$. Because $L\,I(\delta)^{-1}x(t) = (L^1\,I_1(\delta^1)^{-1}x_1(\tau_1))\otimes(L^2\,I_2(\delta^2)^{-1}x_2(\tau_2))$ with $I_i(\delta^i) = \int x_i(\tau)\,x_i(\tau)'\delta^i(d^-)$ the score function (2.6) can be easily calculated. Moreover, if $N = 16\,K$ with $K\in\mathbb{N}$ and d has the same weights and support points as δ, then $\mathcal{N}(d) = 15\,K$.

3 Robust and efficient testing

The classical F-test statistic for testing $H_0 : L\beta = l$ has the form

$$T_{LS}(y,d) = N\,(\widehat{\varphi}_{LS}(y,d) - l)' C_{LS}(y,d)^{-1}\,(\widehat{\varphi}_{LS}(y,d) - l),$$

where C_{LS} is the covariance matrix of the Gauss-Markov estimator $\widehat{\varphi}_{LS}$ for $L\beta$. If we replace the Gauss-Markov estimator by a robust one-step M-estimator $\widehat{\varphi}_\psi$ then the test statistic is more robust. Thereby, C_{LS} should be replaced by $C(\psi,\delta)$, the asymptotic covariance matrix of the one-step M-estimator. Hence, a test statistic based on a one-step M-estimator has the form

$$T_\psi(y,d) = N\,(\widehat{\varphi}_\psi(y,d) - l)' C(\psi,\delta)^{-1}\,(\widehat{\varphi}_\psi(y,d) - l).$$

The breakdown point of T_ψ is equal or higher than the breakdown point of the underlying one-step M-estimator $\widehat{\varphi}_\psi$. Moreover, T_ψ has asymptotically a chi-squared distribution from which asymptotic robustness and asymptotic efficiency properties can be obtained. It can be shown that the maximum asymptotic bias of the level caused by contaminated errors of the form (2.5) is a strictly monotone function of

$$\|\psi' C(\psi,\delta)^{-1}\psi\|_\delta := \operatorname{ess}\sup_{n_{(0,1)}\otimes\delta} |\psi(z,t)' C(\psi,\delta)^{-1}\psi(z,t)|.$$

The efficiency at the ideal normal model is maximized if $C(\psi,\delta)$ is minimized. See Müller (1992a,b), Rieder (1994, Section 5.4), Heritier and Ronchetti (1994). In general situations most efficient robust tests, i.e. tests which minimize the determinant of $C(\psi,\delta)$ under the robustness constraint $\|\psi' C(\psi,\delta)^{-1}\psi\|_\delta \leq b$, are not available. But if δ is a D-optimal design, i.e.

$$\delta \in \arg\min_{\tilde{\delta}} \det(L\,I(\tilde{\delta})^- L'),$$

then the score function ψ_b of the most efficient robust test has the form

$$\psi_b(z,t) = L\,I(\delta)^- x(t)\,\mathrm{sign}(z) \frac{\min(|z|, \sqrt{b}\,w_b)}{2\Phi(\sqrt{b}\,w_b) - 1}, \tag{3.7}$$

where $w_b^2 = \frac{1}{s} g(\sqrt{b}\, w_b) > 0$ and $g(w) = \int \min(|z|, w)^2\, n_{(0,1)}(dz)$. The fixed point w_b can be easily calculated by Newton's method. See Müller (1995b, 1995c, 1997). It is also no problem to obtain the D-optimal designs for several multi-factor models as for models with all, no or complete M-factors interactions. For that the D-optimal designs are products of the D-optimal designs for the single factor models (see Schwabe 1996).

Similarly as for estimation, at a design d which approaches a D-optimal design the underlying one-step M-estimator of the tests should have the following parts: ψ_b of (3.7) is the score function. If $\mathcal{N}(d)$ can be calculated, then the h-trimmed L_2 estimator with $[\frac{N+\mathcal{N}(d)+1}{2}] \leq h \leq [\frac{N+\mathcal{N}(d)+2}{2}]$ should be used as initial estimator. If $\mathcal{N}(d)$ is difficult to calculate, then the Cauchy estimator can be used as initial estimator.

Example: Quadratic regression. Assume $x(t)$, β, Δ, $x_i(\tau_i)$, β^i and Δ^i, $i = 1, 2$, as in the above example. Let be $L^i = \begin{pmatrix} 0 & 1 & 0 \\ 0 & 0 & 1 \end{pmatrix}$ and $L = L^1 \otimes L^2$. Then $L\beta = (\beta_4, \beta_5, \beta_7, \beta_8)'$, i.e. all interaction terms are of interest. Hence, $H_0 : L\beta = 0$ is the hypothesis that there are no interactions. The D-optimal designs for $L^i\beta^i = (\beta_1^i, \beta_2^i)'$ are $\delta^i = \frac{1}{3}(e_{-1} + e_0 + e_1)$ so that according to Corollary 4.6 in Schwabe (1996) $\delta = \delta^1 \otimes \delta^2$, i.e. the design which assigns weights $\frac{1}{9}$ to each point of $\{-1, 0, 1\} \times \{-1, 0, 1\}$, is D-optimal for $L\beta$ in Δ. As for the estimation example we have $L\, I(\delta)^{-1} x(t) = (L^1\, I_1(\delta^1)^{-1} x_1(\tau_1)) \otimes (L^2\, I_2(\delta^2)^{-1} x_2(\tau_2))$ so that the score function (3.7) can be easily calculated. Moreover, if $N = 9\, K$ with $K \in \mathbb{N}$ and d has the same weights and support points as δ, then $\mathcal{N}(d) = 8\, K$.

References

Coakley, C.W. and Mili, L. (1993). Exact fit points under simple regression with replication. *Statist. Probab. Lett.* **17**, 265-271.

Ellis, S. P. and Morgenthaler, S. (1992). Leverage and breakdown in L_1 regression. *J. Amer. Statist. Assoc.* **87**, 143–148.

Hampel, F. R., Ronchetti, E. M., Rousseeuw, P. J. and Stahel, W. A. (1986). *Robust Statistics—The Approach Based on Influence Functions.* Wiley, New York.

He, X., Jurečková, J., Koenker, R. and Portnoy, S. (1990). Tail behavior of regression estimators and their breakdown points. *Econometrica* 58, 1195–1214.

Heritier, S. and Ronchetti, E. (1994). Robust bounded-influence tests in general parametric models. *J. Amer. Statist. Assoc.* **89**, 897-904.

Huber, P. J. (1981). *Robust Statistics.* Wiley, New York.

Jurečková, J. and Portnoy, S. (1987). Asymptotics for one-step M-estimators in regression with application to combining efficiency and high breakdown point. *Commun. Statist.-Theory Meth.* **16**, 2187-2199.

Kurotschka, V. and Müller, Ch.H. (1992). Optimum robust estimation of linear aspects in conditionally contaminated linear models. *Ann. Statist.* **20**, 331-350.

Mili, L. and Coakley, C. W. (1993). Robust estimation in structured linear regression. *Technical Report No. 93-13*, Department of Statistics, Virginia Polytechnic Institute and State University. (To appear in Ann. Statist.)

Mizera, I. and Müller, Ch.H. (1996). Breakdown points and variation exponents of robust M-estimators in linear models. *Preprint No. A-22-96*, Freie Universität Berlin, Fachbereich Mathematik und Informatik.

Müller, Ch.H. (1992a). L_1-estimation and testing in conditionally contaminated linear models. In *L_1-Statistical Analysis and Related Methods*, ed. Y. Dodge, North-Holland, Amsterdam, 69-76.

Müller, Ch.H. (1992b). Robust testing in conditionally contaminated linear models. *Preprint No. A-92-12*, Freie Universität Berlin, Fachbereich Mathematik und Informatik.

Müller, Ch.H. (1994a). Optimal designs for robust estimation in conditionally contaminated linear models. *J. Statist. Plann. Inference.* **38**, 125-140.

Müller, Ch.H. (1994b). One-step-M-estimators in conditionally contaminated linear models. *Stat. Decis.* **12**, 331-342.

Müller, Ch.H. (1995a). Breakdown points for designed experiments. *J. Statist. Plann. Inference.* **45**, 413-427.

Müller, Ch.H. (1995b). Outlier Robust Inference for Planned Experiments. *Habilitationsschrift.* Freie Universität Berlin.

Müller, Ch.H. (1995c). Optimum robust testing in linear models. *Preprint No. A-28-95*, Freie Universität Berlin, Fachbereich Mathematik und Informatik.

Müller, Ch.H. (1996). Computing high breakdown point estimators for planned experiments and for models with qualitative factors. In *COMPSTAT'96, Proceedings in Computational Statistics,* ed. A. Prat, Physica-Verlag, 379-384.

Müller, Ch. H. (1997). Robust Planning and Analysis of Experiments. *To appear as Springer Lecture Notes in Statistics.*

Rousseeuw, P.J. and Croux, C. (1993). Alternatives to the median absolute deviation. *J. Amer. Statist. Assoc.* **88**, 1273-1283.

Schwabe, R. (1996). *Optimum Designs for Multi-Factor Models.* Springer, New York.

Construction of Some New Orthogonal Main-Effect Designs

CHRISTOS KOUKOUVINOS
Department of Mathematics, National Technical University of Athens, Zografou 15773, Athens, Greece.

Key words:
Autocorrelation, orthogonal design, sequence, linear model.

Abstract:
Orthogonal designs are used widely in many experiments. In the present paper we give a method for construction of orthogonal main–effect factorial designs. The method is based on sequences with entries from $\{0, \pm a, \pm b, \pm c\}$ on the commuting variables a, b, c and zero autocorrelation function. In the sequel we construct some new orthogonal factorial designs.

1 Introduction

Consider an experimental situation in which a response y depends on k explanatory variables $x_1, \ldots, x_k$ with the first order relationship of the form

$$\mathbf{y} = X\mathbf{b} + \mathbf{e}$$

where $\mathbf{y}$ is an $n \times 1$ vector of observations, the design matrix X is $n \times k$ whose jth row is of the form $(x_{j1}, x_{j2}, \ldots, x_{jk})$, $j = 1, 2, \ldots, n$, $\mathbf{b}$ is the $k \times 1$ vector of coefficients to be estimated, and $\mathbf{e}$ is the $n \times 1$ vector of errors. We assume that $\mathbf{e}$ is a random vector distributed with mean 0 and covariance matrix $\sigma^2 I_n$. For a given matrix X of rank k, an unbiased estimator of the parameter $\mathbf{b}$ based on the response $\mathbf{y}$ is the ordinary least square estimator

$$\hat{\mathbf{b}} = (X^T X)^{-1} X^T \mathbf{y}$$

It is well known that

$$E(\hat{\mathbf{b}}) = \mathbf{b} \text{ and } Var\,(\hat{\mathbf{b}}) = \sigma^2 (X^T X)^{-1}$$

Furthermore, $\hat{\mathbf{b}}$ has the least covariance matrix among all linear unbiased estimators of $\mathbf{b}$. Hence $\hat{\mathbf{b}}$ is called the *best linear unbiased estimator* of $\mathbf{b}$.

*Kitsos, C.P. and Edler, L., Eds.:Industrial Statistics, Physica-Verlag, Heidelberg, 1997

With the additional hypothesis of the normality of $\mathbf{e}$, i.e., $\mathbf{e} \sim N(\mathbf{0}, \sigma^2 I_n)$, the estimator $\widehat{\mathbf{b}}$ satisfies also the maximum likelihood criterion. The $k \times k$ information matrix $X^T X$ is taken to be full rank. The variance of predicted response at $\mathbf{x}$ is

$$Var\,(\widehat{y}(\mathbf{x})) = \sigma^2 \mathbf{x}^T (X^T X)^{-1} \mathbf{x}$$

Optimum design theory (Kiefer [5, 6]) is concerned with the choice of design matrix X from a given experimental region to minimize various functions of the matrix valued variance of the ordinary least square estimator of $\mathbf{b}$, where the various functions are proposed by practical consideration. This optimal design is one of many optimal regression designs. Since the information matrix about $\mathbf{b}$ from the experimental region is proportional to $X^T X$, the problem for finding an optimal design is equivalent to choosing the design matrix X to make the matrix $X^T X$ large, in some sense. That is, find an $X \in \mathcal{X}$ which maximizes some real value functions $\Phi(X^T X)$, where $\mathcal{X}$ is an allowable (or feasible) region of design matrices and Φ is a given function that is proposed by practical consideration. Different function Φ gives various criteria of optimal design. This criterion has been called Φ–optimality. Let Φ be the determinant of a matrix, i.e.,

$$\Phi(X^T X) = det\,(X^T X)$$

this criterion is called D–optimality (Kiefer [5]). If the error $\mathbf{e}$ is distributed normally, for a given confidence coefficient and a residual sum of squares, the content of a confidence ellipsoid

$$\left\{ \mathbf{b} : (\mathbf{b} - \widehat{\mathbf{b}})^T X^T X (\mathbf{b} - \widehat{\mathbf{b}}) \leq c \right\}$$

for $\mathbf{b}$ is proportional to the reciprocal of the square root of $det\,(X^T X)$, where c is a constant. It is a natural design criterion that makes the ellipsoid small, i.e., maximizing $det\,(X^T X)$. This is the reason why we consider the D–optimality.
A further property of the D–optimal designs, in the linear case, is the invariance under right linear transformation. Let X be a real $n \times k$ matrix, the design of a linear model, and A an orthogonal $k \times k$ matrix. The transformation $D = XA$ does not modify the D–optimal property. Indeed, it is easy to see that the determinant of the matrix $D^T D$ is

$$det\,(D^T D) = (det\,(A))^2 det\,(X^T X) = det\,(X^T X)$$

Factorial designs are widely used in experiments involving several factors where it is necessary to study the joint effect of the factors on a response. There are several special cases of the general factorial design that are important because they are widely used in research work and also because they form the basis of other designs of considerable practical value. The most important of these special cases is that of k factors, each at two or three levels. In this case, if $X = (x_{ij})$, $i = 1, \ldots, n$, $j = 1, \ldots, k$ is the design matrix, each x_{ij} can be coded as ± 1 or $0, \pm 1$. The design is then determined by the $n \times k$ matrix of elements ± 1 or $0, \pm 1$. These designs are studied in detail by Hotelling [4], Mood [9], Raghavarao [10], and Banerjee [1]. Factorial designs of resolution III are such that all main effects are estimable, ignoring two-factor interactions and all higher order interactions. If the design permits estimation of all main effects without correlation, the design is said to be an *orthogonal main-effect design*. Several combinatorial designs have been found to be extremely useful in constructing orthogonal main-effect designs. In the present paper we study some of these designs which are saturated and provide orthogonal estimates of all main effects when all interactions are assumed to be zero, and we construct some new orthogonal factorial designs. Our construction is based on sequences having as entries commuting variables and zero autocorrelation function.

2 Orthogonal designs and their construction

An *orthogonal design* of order n and type $(s_1, s_2, \ldots, s_u)$ $(s_i > 0)$, denoted $OD(n; s_1, s_2, \ldots, s_u)$, on the commuting variables $x_1, x_2, \ldots, x_u$ is an $n \times n$ matrix A with entries from $\{0, \pm x_1, \pm x_2, \ldots, \pm x_u\}$ such that

$$AA^T = (\sum_{i=1}^{u} s_i x_i^2) I_n$$

Alternatively, the rows of A are formally orthogonal and each row has precisely s_i entries of the type $\pm x_i$. In [2], where this was first defined, it was mentioned that

$$A^T A = (\sum_{i=1}^{u} s_i x_i^2) I_n$$

and so our alternative description of A applies equally well to the columns of A. It was also shown in [2] that $u \leq \rho(n)$, where $\rho(n)$ (Radon's function) is defined by $\rho(n) = 8c + 2^d$, when $n = 2^a b$, b odd, $a = 4c + d$, $0 \leq d < 4$.

A weighing matrix $W = W(n, k)$ is a square matrix with entries $0, \pm 1$ having k non-zero entries per row and column and inner product of distinct rows zero. Hence W satisfies $WW^T = kI_n$, and W is equivalent to an orthogonal design $OD(n; k)$. The number k is called the *weight* of W.

If $k = n$, that is, all the entries of W are ± 1 and $WW^T = nI_n$, then W is called an Hadamard matrix of order n. In this case $n = 1, 2$ or $n \equiv 0 (mod\ 4)$.

Given the sequence $A = \{a_1, a_2, \ldots, a_n\}$ of length n the *non-periodic autocorrelation function* $N_A(s)$ is defined as

$$N_A(s) = \sum_{i=1}^{n-s} a_i a_{i+s}, \quad s = 0, 1, \ldots, n-1, \tag{1}$$

Given A as above of length n the *periodic autocorrelation function* $P_A(s)$ is defined, reducing $i + s$ modulo n, as

$$P_A(s) = \sum_{i=1}^{n} a_i a_{i+s}, \quad s = 0, 1, \ldots, n-1. \tag{2}$$

The following theorem which uses four circulant matrices is very useful in our construction for orthogonal designs.

Theorem 1 *[3, Theorem 4.49] Suppose there exist four circulant matrices A, B, C, D of order n satisfying*

$$AA^T + BB^T + CC^T + DD^T = fI_n$$

Let R be the back diagonal matrix. Then

$$GS = \begin{pmatrix} A & BR & CR & DR \\ -BR & A & D^TR & -C^TR \\ -CR & -D^TR & A & B^TR \\ -DR & C^TR & -B^TR & A \end{pmatrix}$$

is a $W(4n, f)$ when A, B, C, D are $(0, 1, -1)$ matrices, and an orthogonal design $OD(4n; s_1, s_2, \ldots, s_u)$ on $x_1, x_2, \ldots, x_u$ when A, B, C, D have entries from $\{0, \pm x_1, \ldots, \pm x_u\}$ and $f = \sum_{j=1}^{u}(s_j x_j^2)$. □

Corollary 1 *If there are four sequences A, B, C, D of length n with entries from $\{0, \pm x_1, \pm x_2, \pm x_3, \pm x_4\}$ with zero periodic or non-periodic autocorrelation function, then these sequnces can be used as the first rows of circulant matrices which can be used in the Goethals-Seidel array to form an $OD(4n; s_1, s_2, s_3, s_4)$. We note that if there are sequences of length n with zero non-periodic autocorrelation function, then there are sequences of length $n + m$ for all $m \geq 0$.* □

The above method for constructing orthogonal designs has been used by Koukouvinos [7, 8].

3 New orthogonal designs and sequences with zero autocorrelation function

In this section we give new sequences with entries from $\{0, \pm a, \pm b, \pm c\}$ on the commuting variables a, b, c and zero autocorrelation function. Then we use these sequences to construct some new orthogonal designs. If the variables a, b, c take values from the set $\{0, 1, -1\}$, then we produce some new two or three level orthogonal factorial designs.

Design	Sequences
$OD(36;3,3,24)$	$A_1 = (c\ \ c\ \ c\ \ -a\ \ 0\ \ -b\ \ -c\ \ c\ \ -c)$
	$A_2 = (c\ \ c\ \ -c\ \ -a\ \ 0\ \ b\ \ -c\ \ c\ \ c)$
	$A_3 = (c\ \ c\ \ c\ \ 0\ \ a\ \ 0\ \ c\ \ -c\ \ c)$
	$A_4 = (c\ \ c\ \ -c\ \ 0\ \ -b\ \ 0\ \ c\ \ -c\ \ -c)$
$OD(36;3,6,17)$	$A_1 = (c\ \ b\ \ c\ \ 0\ \ a\ \ b\ \ -c\ \ 0\ \ -c)$
	$A_2 = (a\ \ -b\ \ c\ \ -b\ \ c\ \ b\ \ 0\ \ 0\ \ 0)$
	$A_3 = (b\ \ c\ \ 0\ \ 0\ \ 0\ \ -a\ \ 0\ \ c\ \ 0)$
	$A_4 = (c\ \ c\ \ c\ \ -c\ \ -c\ \ c\ \ c\ \ -c\ \ c)$
$OD(36;3,6,24)$	$A_1 = (c\ \ c\ \ -c\ \ 0\ \ -a\ \ -b\ \ c\ \ -c\ \ c)$
	$A_2 = (c\ \ -c\ \ -c\ \ b\ \ 0\ \ -a\ \ c\ \ c\ \ c)$
	$A_3 = (c\ \ c\ \ c\ \ b\ \ -b\ \ a\ \ c\ \ c\ \ -c)$
	$A_4 = (c\ \ -c\ \ c\ \ 0\ \ b\ \ b\ \ c\ \ -c\ \ -c)$
$OD(36;4,5,20)$	$A_1 = (c\ \ 0\ \ a\ \ -b\ \ a\ \ b\ \ -c\ \ 0\ \ 0)$
	$A_2 = (c\ \ 0\ \ a\ \ -b\ \ -a\ \ -b\ \ c\ \ 0\ \ 0)$
	$A_3 = (c\ \ c\ \ -c\ \ c\ \ b\ \ c\ \ c\ \ c\ \ -c)$
	$A_4 = (c\ \ c\ \ -c\ \ c\ \ 0\ \ -c\ \ -c\ \ -c\ \ c)$
$OD(36;4,5,25)$	$A_1 = (c\ \ b\ \ 0\ \ a\ \ -c\ \ a\ \ c\ \ -b\ \ -c)$
	$A_2 = (c\ \ b\ \ 0\ \ a\ \ -c\ \ -a\ \ -c\ \ b\ \ c)$
	$A_3 = (c\ \ c\ \ c\ \ -c\ \ b\ \ -c\ \ -c\ \ c\ \ -c)$
	$A_4 = (c\ \ -c\ \ c\ \ c\ \ c\ \ -c\ \ c\ \ c\ \ c)$
$OD(36;6,12,16)$	$A_1 = (a\ \ c\ \ b\ \ -c\ \ b\ \ b\ \ c\ \ -a\ \ c)$
	$A_2 = (a\ \ c\ \ b\ \ -c\ \ b\ \ -b\ \ -c\ \ a\ \ -c)$
	$A_3 = (a\ \ c\ \ 0\ \ c\ \ -b\ \ b\ \ c\ \ -b\ \ -c)$
	$A_4 = (a\ \ c\ \ 0\ \ c\ \ -b\ \ -b\ \ -c\ \ b\ \ c)$

Table 1: Sequences of length 9 with zero non-periodic autocorrelation function

Theorem 2 *There exist orthogonal designs*

$$OD\,(4n;3,3,24),\ OD\,(4n;3,6,17),\ OD\,(4n;3,6,24),$$
$$OD\,(4n;4,5,20),\ OD\,(4n;4,5,25),\ OD\,(4n;6,12,16)$$

for all $n \geq 9$, constructed using four circulant matrices in the Goethals-Seidel array.

Proof. We use the sequences given in the Table 1, which have zero non-periodic autocorrelation function, as the first rows of the corresponding circulant matrices in the Goethals-Seidel array to obtain the required orthogonal designs. Since these sequences have zero non-periodic autocorrelation function, the sequences are first padded with sufficient zeros added to the end to make their length $n \geq 9$. □

Design	Sequences
$OD(36;3,7,23)$	$A_1 = (\ a\quad b\quad c\quad -c\quad -c\quad c\quad b\quad -c\quad 0)$
	$A_2 = (\ b\quad a\quad -b\quad c\quad -b\quad -c\quad c\quad c\quad c)$
	$A_3 = (\ a\quad -b\quad -c\quad -c\quad c\quad -c\quad c\quad -c\quad 0)$
	$A_4 = (\ c\quad -b\quad c\quad -c\quad -c\quad -c\quad -c\quad -c\quad 0)$
$OD(36;3,8,22)$	$A_1 = (\ a\quad b\quad b\quad -c\quad c\quad c\quad -c\quad -c\quad 0)$
	$A_2 = (\ a\quad -b\quad -c\quad c\quad -c\quad -c\quad -b\quad c\quad 0)$
	$A_3 = (\ a\quad -c\quad -b\quad b\quad -c\quad c\quad c\quad c\quad c)$
	$A_4 = (\ c\quad c\quad -b\quad c\quad -c\quad c\quad b\quad c\quad 0)$
$OD(36;3,9,18)$	$A_1 = (\ a\quad b\quad c\quad b\quad -c\quad -c\quad c\quad -c\quad -c)$
	$A_2 = (\ c\quad -b\quad a\quad -c\quad -c\quad -b\quad -c\quad c\quad 0)$
	$A_3 = (\ c\quad a\quad c\quad 0\quad -c\quad c\quad c\quad 0\quad 0)$
	$A_4 = (\ b\quad b\quad -b\quad c\quad -b\quad c\quad b\quad 0\quad 0)$
$OD(36;5,8,23)$	$A_1 = (\ a\quad a\quad c\quad -c\quad c\quad c\quad -c\quad -c\quad -c)$
	$A_2 = (\ a\quad -a\quad -c\quad -c\quad -c\quad -a\quad c\quad c\quad -c)$
	$A_3 = (\ b\quad -b\quad -b\quad -c\quad -b\quad -c\quad c\quad -c\quad -c)$
	$A_4 = (\ b\quad -b\quad c\quad -c\quad c\quad -b\quad -b\quad c\quad c)$
$OD(36;5,13,18)$	$A_1 = (\ a\quad c\quad a\quad -c\quad -b\quad -b\quad b\quad b\quad -c)$
	$A_2 = (\ a\quad -a\quad -a\quad -c\quad c\quad -c\quad c\quad -c\quad -c)$
	$A_3 = (\ b\quad b\quad c\quad c\quad -c\quad -c\quad b\quad -c\quad -c)$
	$A_4 = (\ b\quad -b\quad b\quad -b\quad -b\quad -c\quad -b\quad -c\quad -c)$
$OD(36;6,7,23)$	$A_1 = (\ a\quad a\quad -b\quad c\quad -c\quad c\quad -c\quad c\quad c)$
	$A_2 = (\ a\quad -a\quad b\quad c\quad -b\quad -a\quad -b\quad c\quad c)$
	$A_3 = (\ a\quad -c\quad c\quad b\quad -c\quad c\quad c\quad -c\quad -c)$
	$A_4 = (\ b\quad -c\quad -c\quad c\quad b\quad c\quad c\quad c\quad c)$

Table 2 : Sequences of length 9 with zero periodic autocorrelation function

Theorem 3 *There exist orthogonal designs*

$$OD\,(36;3,7,23),\ OD\,(36;3,8,22),\ OD\,(36;3,9,18),$$
$$OD\,(36;5,8,23),\ OD\,(36;5,13,18)\ ,\ OD\,(36;6,7,23)$$

constructed using four circulant matrices in the Goethals-Seidel array.

Proof. We use the sequences given in the Table 2, which have zero periodic autocorrelation function, as the first rows of the corresponding circulant matrices in the Goethals-Seidel array to obtain the required orthogonal designs. □

References

[1] K.S.Banerjee, *Weighing designs for Chemistry, Medicine, Economics, Operations Research and Statistics*, Marcel Dekker, New York, 1975.

[2] A.V.Geramita, J.M.Geramita, and J.Seberry Wallis, Orthogonal designs, *Linear and Multilinear Algebra*, 3, (1976), 281-306.

[3] A.V.Geramita, and J.Seberry, *Orthogonal designs: Quadratic forms and Hadamard matrices*, Marcel Dekker, New York-Basel, 1979.

[4] H.Hotelling, Some improvements in weighing and other experimental techniques, *Ann. Math. Stat.*, 15, (1944), 297-306.

[5] J.Kiefer, On non-randomized optimality and randomized non-optimality of symmetrical designs, *Ann. Math. Stat.*, 29, (1958), 675-699.

[6] J.Kiefer, Optimum experimental designs, *J. Roy. Statist. Soc. Ser. B*, 21, (1959), 272-319.

[7] C.Koukouvinos, Orthogonal 2^k and 3^k factorial designs constructed using sequences with zero autocorrelation, *Statistics and Probability Letters*, 28, (1996), 59-63.

[8] C.Koukouvinos, Some new orthogonal designs in linear regression models, *Statistics and Probability Letters*, 29, (1996), 125-129.

[9] A.Mood, On Hotelling's weighing problem, *Ann. Math. Statis.*, 17, (1946), 432-446.

[10] D.Raghavarao, *Constructions and Combinatorial Problems in Design of Experiments*, J. Wiley and Sons, New York, 1971.

Extended V–robustness for Two-level Orthogonal Resolution V Designs

CINZIA MORTARINO AND LUIGI SALMASO
Università degli Studi di Padova
Dipartimento di Scienze Statistiche
Via S. Francesco, 33 – 35121 Padova – Italia
e-mail: mortarin@hal.stat.unipd.it
lsalmaso@hal.stat.unipd.it

Key Words: BLU estimators, Factorial designs, Irregular fractions, Multiresponse designs

Abstract

Extended V–robustness property allows to prove *coincidence* between OLSE and WLSE for parameters of a multiresponse linear model when a regular fraction of a 2^k factorial design is performed (GUSEO and MORTARINO (1997)). With a similar assumption, the same result was also proved for Addelman's non-orthogonal irregular fractions (MORTARINO (1996)).

The aim of this paper is to verify if a similar result, with regard to two-level *orthogonal resolution V designs* (SALMASO (1996), KOUNIAS and SALMASO (1997)), holds.

A canonical assumption of extended V–robust covariance pattern is not sensible here, due to the presence of *repeated* runs. Hence, the original assumption is modified and this leads to the coincidence between OLSE and WLSE for designs with 48 and 96 runs. If 80 and 112 run designs are considered, a different approach is needed.

1 Introduction

Canonical assumptions supporting standard procedures are often too strong to hold for real experiments and, mostly, this heavily affects their performance.

*Kitsos, C.P. and Edler, L., Eds.:Industrial Statistics, Physica-Verlag, Heidelberg, 1997

Nevertheless, there are situations where these assumptions can be violated (at least up to a certain degree) without loss of efficiency. Since it is not always easy to recognize if all assumptions hold for a certain situation, it is important to investigate the robustness of standard procedures.

The aim of this paper is to study the effect on the efficiency of OLSE for parameters of a linear model when an orthogonal resolution V design (SALMASO (1996), KOUNIAS and SALMASO (1997)) is performed and a certain correlation pattern is present among different runs. This correlation pattern is obtained as a more general version of the one introduced by KROUSE (1994) for the uniresponse case and by GUSEO and MORTARINO (1997) for the multiresponse case. The construction of the class of designs here considered relies on the assumption that only main effects and first order interactions have significative size and thus higher order interactions can be neglected. These designs are flexible towards the choice of the number of observations, since we consider irregular fractions of the full two-level factorial design and we examine designs with N = 48, 80, 96, 112.

In section 2. we introduce the model used throughout the paper, with a description of symbols and notations and we present the basic ideas of extended V–robust covariance patterns. In sections 3. and 4. the main question is examined. In section 3. we examine designs with N = 48, 96 runs: a profitable configuration among runs is given for each design matrix and we prove the coincidence between OLSE and WLSE for the parameters of a linear model (both uniresponse and multiresponse) when the assumption of extended V-robust covariance pattern holds. For N = 80, 112, the question is further analyzed in section 4. since problems arise from the configuration of these designs. With some restrictions on the designs, we are able to find a n.s. condition to prove coincidence between OLSE and WLSE also for N = 80, 112 runs.

2 V–robust covariance patterns

Let

$$y_{iu} = \boldsymbol{z}'_{iu}\boldsymbol{\beta}_i + \varepsilon_{iu}, \qquad i = 1, 2, \ldots, h, \quad u = 1, 2, \ldots, N, \tag{2.1}$$

be the multiresponse model, where N is the number of experimental units, h is the number of responses observed on each unit, y_{iu} represents the i-th response corresponding to the u-th unit, $\boldsymbol{z}'_{iu}$ is the u-th row of the matrix $\boldsymbol{Z}_i$: this row contains the levels of independent factors corresponding to the u-th unit and expected to influence the i–th response; ε_{iu} denotes the zero mean random error corresponding to y_{iu}. In the following, it will be assumed that for every i, $\boldsymbol{Z}_i \equiv \boldsymbol{Z}_{N\times p}$.

The matrix form of model (2.1) will be useful:

$$\boldsymbol{y}^*_{Nh\times 1} = \boldsymbol{Z}^*_{Nh\times ph}\,\boldsymbol{\beta}^*_{ph\times 1} + \boldsymbol{\varepsilon}^*_{Nh\times 1}\,; \tag{2.2}$$

this is obtained by stacking the N vectors corresponding to the h responses measured on each unit. The vector $\boldsymbol{\beta}^*$ is obtained through an appropriate rearrangement of the elements of the $\boldsymbol{\beta}_i$'s. Finally, let

$$\boldsymbol{\varepsilon}_u = [\varepsilon_{1u}, \varepsilon_{2u}, \ldots, \varepsilon_{hu}]'$$

be the vector of errors corresponding to the outcomes on the u-th unit. Let $\boldsymbol{\Sigma}^*$ be the covariance matrix of vector

$$\boldsymbol{\varepsilon}^* = [\boldsymbol{\varepsilon}_1', \boldsymbol{\varepsilon}_2', \ldots \boldsymbol{\varepsilon}_N']' .$$

This covariance matrix is supposed to belong to a special class introduced by GUSEO and MORTARINO (1997) as an extension of a previous one given by KROUSE (1994) for the uniresponse case. In GUSEO and MORTARINO (1997) an interpretation of both definitions is proposed together with some examples of situations satisfying these assumptions.

Definition 2.1. *(extended V–robust matrix) A covariance matrix* $\boldsymbol{\Sigma}^*_{Nh\times Nh}$, *partitioned in* N^2 *blocks of size* $h \times h$ *is extended* V*–robust with respect to*

$$\boldsymbol{M}^*_{Nh\times t} = [\,\boldsymbol{M}_1^{*\prime} \quad \boldsymbol{M}_2^{*\prime} \quad \ldots \quad \boldsymbol{M}_N^{*\prime}\,]' ,$$

where $\boldsymbol{M}^*_j$ *is for every* j, *a* $(h \times t)$ *submatrix, if* $\forall u, w = 1, 2, \ldots, N$,

$$\boldsymbol{\Sigma}^*_{uw} = \mathrm{Cov}(\boldsymbol{\varepsilon}_u, \boldsymbol{\varepsilon}_w) = g(\boldsymbol{M}^*_u \odot \boldsymbol{M}^*_w) ,$$

where $g : \mathbb{R}^{ht} \to C_h$, C_h *denotes the space of all* $h \times h$ *matrices, and* g *is such that* $\boldsymbol{\Sigma}^*$ *is positive definite.*

These definitions were introduced because of the possibility to prove optimality of OLSE for parameters of model (2.2) when a regular fraction of a two level factorial design is performed. That result was proved by KROUSE (1994) for the uniresponse case and extended by GUSEO and MORTARINO (1997) to the multiresponse case. In particular, for model (2.2), when $\boldsymbol{Z}$ is the matrix associated to a 2^{k-n} design and $V(\boldsymbol{\varepsilon}^*) = \boldsymbol{\Sigma}^*$ is an extended V–robust covariance matrix with respect to $\boldsymbol{Z}^* = \boldsymbol{Z} \otimes \boldsymbol{I}_h$, it was proved that

$$\boldsymbol{\Sigma}^*_{Nh\times Nh}\, \boldsymbol{Z}^*_{Nh\times ph} = \boldsymbol{Z}^*_{Nh\times ph}\, \boldsymbol{B}^*_{ph\times ph} , \tag{2.3}$$

where $\boldsymbol{B}^*$ is a block–diagonal matrix. Formula (2.3) is exactly one of the conditions proved by RAO (1967) and ZYSKIND (1967) leading to the coincidence $\hat{\boldsymbol{\beta}}^*_{\mathrm{WLS}} = \tilde{\boldsymbol{\beta}}^*_{\mathrm{OLS}}$.

A further extension was also proved (MORTARINO (1996)) in order to deal with some irregular non-orthogonal fractions described in detail by ADDELMAN (1961).

3 V–robustness for orthogonal designs with N =48, 96 runs

The aim of this paper is to see if the results mentioned in section 2. can be extended to *two level orthognal resolution V designs* described in SALMASO (1996) and KOUNIAS and SALMASO (1997). Since *regular* fractions belong to the class of designs already treated in GUSEO and MORTARINO (1997), we are going to examine only *irregular* fractions. In particular, we start by investigating the 48-run design and then the 96-run design, which is similar to the first one. In order to make the description clearer, at first we will analyze the uniresponse case ($h = 1$).

48-run designs. The K-algorithm (SALMASO (1996)) provides the maximum number, k, of orthogonal factors for each $N \equiv 0 \bmod 16$, to form a resolution V design. For $N = 48$ the maximum number of factors to form a resolution V design is five and we consider here the design matrix:

$$\boldsymbol{H}_{48\times 5} = \mathbf{1}_3 \otimes [\boldsymbol{H}_{16} \ + \boldsymbol{e}],$$

where $\boldsymbol{H}_{16}$ is the (16×4) design matrix of the 2^4 resolution V design with four main factors and $\boldsymbol{e}$ is the product of the previous four main factors, i.e. the fifth factor is obtained from the defining contrast $E = ABCD$.

Let

$$\boldsymbol{y}_{48\times 1} = \boldsymbol{Z}_{48\times 16}\boldsymbol{\beta}_{16\times 1} + \boldsymbol{\varepsilon}_{48\times 1}, \tag{3.1}$$

be the uniresponse model to be analyzed; note that the model matrix is structured as follows

$$\boldsymbol{Z} = \mathbf{1}_3 \otimes \boldsymbol{X}_{16},$$

where $\boldsymbol{X}_{16} = [\mathbf{1}, \ \boldsymbol{H}_{16}, \ \boldsymbol{e}, \ \boldsymbol{P}^{\cdot}$; here $\mathbf{1}$ is a 16-vector with all elements equal to 1, while $\boldsymbol{P}$ is a (16×10) matrix whose columns are direct products between couples of columns of $[\boldsymbol{H}, \ \boldsymbol{e}]$ (these products represent first order interactions). Finally, let $\boldsymbol{\Sigma}$ denote $\mathrm{Cov}(\boldsymbol{\varepsilon})$.

It is important to investigate the meaning of an assumption of V-robustness of $\boldsymbol{\Sigma}$ w.r.t. $\boldsymbol{Z}$. In this case, due to the structure of $\boldsymbol{Z}$,

$$\sigma_{11} = \sigma_{ii} = \sigma_{i,i+16} = \sigma_{i,i+32} = \sigma_{i+16,i+32} \qquad \text{for } i = 1, 2, \ldots, 16.$$

This would be equivalent to the requirement $\mathrm{Var}(T) = \mathrm{Cov}(T, V)$, where T and V are random variables representing the "error" term associated to <u>different</u> runs, which are, however, realized with the same experimental setup. If we assume that $\mathrm{E}(T) = \mathrm{E}(V) = 0$, then previous condition, for T and V absolutely continuous, for simplicity, with respect to the Lebesgue measure, is equivalent to

$$\int_{\mathcal{T}} t^2 f(t)dt = \int_{\mathcal{T}}\int_{\mathcal{V}} tvf(v|t)f(t)dvdt = \int_{\mathcal{T}} tf(t)\mathrm{E}(V|t)dt \quad \Leftrightarrow \quad t \overset{\text{a.e.}}{=} \mathrm{E}(V|t).$$

If runs with an identical setup were performed simultaneously, then such an assumption might be accepted. It is then obvious that this dependence among variables leads, however, to a singular covariance matrix. In particular, V-robustness property leads to a covariance matrix $\boldsymbol{\Sigma}$ structured as follows:

$$\boldsymbol{\Sigma} = \boldsymbol{J}_3 \otimes \boldsymbol{S}\,, \tag{3.2}$$

where $\boldsymbol{J}_n$ denotes an $n \times n$ matrix having all elements equal to 1 and $\boldsymbol{S}$ is an appropriate 16×16 matrix.

As a consequence of this, it might be sensible to look at a more general covariance pattern (including (3.2) as a special case).

Theorem 3.1. *In the context of model* (3.1), *if* $\boldsymbol{\Sigma}$ *is such that two functions* $f_A(\cdot)$ *and* $f_B(\cdot)$, *mapping* $\mathbb{R}^{16}$ *into* $\mathbb{R}$, *exist so that for* $i,j = 1,2,\ldots,16$

$$\sigma_{i(u),j(u)} = f_A(\boldsymbol{x}'_i \odot \boldsymbol{x}'_j) \qquad \forall u = 1,2,3, \tag{3.3a}$$

$$\sigma_{i(u),j(w)} = f_B(\boldsymbol{x}'_i \odot \boldsymbol{x}'_j) \qquad \forall u,w = 1,2,3 \;\; u \neq w, \tag{3.3b}$$

then

$$\boldsymbol{\Sigma Z} = \boldsymbol{ZG},$$

where $\boldsymbol{G}$ *is an appropriate* 16×16 *matrix. It follows that* $\hat{\boldsymbol{\beta}}_{\mathrm{WLS}} \equiv \tilde{\boldsymbol{\beta}}_{\mathrm{OLS}}$.

For a proof of this theorem see MORTARINO and SALMASO (1996). Remark that assumptions (3.3) lead to a pattern for $\boldsymbol{\Sigma}$ like

$$[\boldsymbol{I}_3 \otimes (\boldsymbol{A} - \boldsymbol{B}) + \boldsymbol{J}_3 \otimes \boldsymbol{B}]\;: \tag{3.4}$$

this pattern equals (3.2) when $\boldsymbol{A} = \boldsymbol{B}$ (i.e. $f_A(\cdot) = f_B(\cdot)$). A brief discussion of the meaning of assumptions (3.3) will be given at the end of this section.

96-run designs. For $N = 96$ we consider an orthogonal resolution V design with six main factors. The model matrix associated to this design is

$$\boldsymbol{Z}_{(96\times 22)} = \boldsymbol{1}_3 \otimes \boldsymbol{W}\,, \tag{3.5}$$

where

$$\boldsymbol{W} = \begin{bmatrix} +1 & \boldsymbol{H}_{16} & +\boldsymbol{e} & +\boldsymbol{e} & \boldsymbol{P}_1 \\ +1 & \boldsymbol{H}_{16} & +\boldsymbol{e} & -\boldsymbol{e} & \boldsymbol{P}_2 \end{bmatrix}$$

and, as before, $\boldsymbol{P}_1$ and $\boldsymbol{P}_2$ are matrices whose columns are obtained as products between couples of previous columns. This partition is useful since we consider three different blocks of runs, each of them with 32 runs. So, with a slight difference in the notation used, let $\sigma_{i(u),j(w)}$ denote the covariance between the i-th run of block u and the j-th run of block w, where $u,w = 1,2,3$ and now $i,j = 1,2,\ldots,32$. It is sensible to use $\boldsymbol{w}'_{i(u)}$ in order to denote the i-th row of block u ($\Rightarrow \boldsymbol{w}'_{i(u)} = \boldsymbol{w}'_t$ for $t = 32(u-1)+i$). With this structure we can prove the following theorem.

Theorem 3.2. *The same result proved in Theorem 3.1. holds for a model based on matrix (3.5), if $\mathbf{\Sigma}$ is such that two functions $f_A(\cdot)$ and $f_B(\cdot)$, mapping $\mathbb{R}^{22}$ into $\mathbb{R}$, exist so that for $i, j = 1, 2, \ldots, 32$*

$$\sigma_{i(u),j(u)} = f_A(\boldsymbol{w}_i' \odot \boldsymbol{w}_j') \quad \forall u = 1,2,3, \tag{3.6a}$$

$$\sigma_{i(u),j(w)} = f_B(\boldsymbol{w}_i' \odot \boldsymbol{w}_j') \quad \forall u, w = 1,2,3 \; u \neq w. \tag{3.6b}$$

The proof of this theorem easily comes as we observe that conditions (3.6) require that

$$\mathbf{\Sigma} = \boldsymbol{I}_3 \otimes (\boldsymbol{A} - \boldsymbol{B}) + \boldsymbol{J}_3 \otimes \boldsymbol{B},$$

where $\boldsymbol{A}$ and $\boldsymbol{B}$ are 32×32 V-robust matrices w.r.t. $\boldsymbol{W}$.

Multiresponse models. Consider model (2.2)

$$\boldsymbol{y}^*_{Nh\times 1} = \boldsymbol{Z}^*_{Nh\times ph}\,\boldsymbol{\beta}^*_{ph\times 1} + \boldsymbol{\varepsilon}^*_{Nh\times 1}, \tag{3.7}$$

where $N = 48$ or $N = 96$ and p is fixed accordingly. The two cases can be treated simultaneously since for both of them the rows of $\boldsymbol{Z}^*$ can be partitioned into three identical groups:

$$\boldsymbol{Z}^*_{48} = \boldsymbol{Z}_{48} \otimes \boldsymbol{I}_h = (\mathbf{1}_3 \otimes \boldsymbol{X}_{16}) \otimes \boldsymbol{I}_h = \mathbf{1}_3 \otimes (\boldsymbol{X}_{16} \otimes \boldsymbol{I}_h)$$

$$\boldsymbol{Z}^*_{96} = \boldsymbol{Z}_{96} \otimes \boldsymbol{I}_h = (\mathbf{1}_3 \otimes \boldsymbol{W}) \otimes \boldsymbol{I}_h = \mathbf{1}_3 \otimes (\boldsymbol{W} \otimes \boldsymbol{I}_h).$$

We now have to introduce an "extension" for extended V–robustness, similar to the one proposed in (3.3) and (3.6) for uniresponse V–robustness. We can then prove the following theorem.

Theorem 3.3. *In the context of model (3.7), with $\boldsymbol{Z}^* = \boldsymbol{Z}^*_{48}$ $[\boldsymbol{Z}^* = \boldsymbol{Z}^*_{96}]$, if $\mathbf{\Sigma}^*$ is such that two functions $\mathcal{F}_A(\cdot)$ and $\mathcal{F}_B(\cdot)$, mapping $\mathbb{R}^{ph}$ into C_h, exist so that for $i, j = 1, 2, \ldots, 16$ $(i, j = 1, 2, \ldots, 32)$*

$$\mathbf{\Sigma}_{i(u),j(u)} = \mathcal{F}_A(\boldsymbol{x}_i' \odot \boldsymbol{x}_j') \quad \forall u = 1,2,3 \tag{3.8a}$$

$$[\,\mathbf{\Sigma}_{i(u),j(u)} = \mathcal{F}_A(\boldsymbol{w}_i' \odot \boldsymbol{w}_j') \quad \forall u = 1,2,3\,] \tag{3.8a'}$$

$$\mathbf{\Sigma}_{i(u),j(w)} = \mathcal{F}_B(\boldsymbol{x}_i' \odot \boldsymbol{x}_j') \quad \forall u = 1,2,3 \; u \neq w \tag{3.8b}$$

$$[\,\mathbf{\Sigma}_{i(u),j(w)} = \mathcal{F}_B(\boldsymbol{w}_i' \odot \boldsymbol{w}_j') \quad \forall u = 1,2,3 \; u \neq w\,], \tag{3.8b'}$$

then

$$\mathbf{\Sigma}^* \boldsymbol{Z}^* = \boldsymbol{Z}^* \boldsymbol{G}^*,$$

where $\boldsymbol{G}^*$ is an appropriate $ph \times ph$ matrix. It follows that $\hat{\boldsymbol{\beta}}^*_{\mathrm{WLS}} \equiv \tilde{\boldsymbol{\beta}}^*_{\mathrm{OLS}}$.

Theorem 3.3. (and its corresponding special cases for uniresponse models included in theorems 3.1. and 3.2.) assume an interesting pattern for the error covariance matrix As for canonical V–robustness, covariances among response components measured on different experimental units are determined by the difference in factors setup of the corresponding runs and, in this way, these covariances can be seen as "distances"among experimental design points (GUSEO and MORTARINO (1997)). Note that because of the generality of functions $\mathcal{F}_A$ and $\mathcal{F}_B$, *no specific patterns* were assumed for covariances among different response components, which are left totally unconstrained.

What is peculiar to assumptions (3.8) (and to (3.3) and (3.6)) is the presence of two independent link functions: experimental units are partitioned into separate blocks of replicates and we have a link functions for "within-block" covariances and another one for "between-blocks" covariances. This means that assumptions (3.8) work in a situation where runs within each block are performed in the same conditions and different replicates (different blocks) can be handled symmetrically. This second requirement excludes, for example, situations where a trend effect in subsequent blocks or a "circular dependence" among blocks is present, but it is satysfied, for example, by many block-sequential experiments.

Observe that model (3.1) with assumptions (3.3) includes a particular variance components model: instead of generalizing V–robust patterns to (3.3) we could have taken into account differences among blocks of runs by introducing in the model a random effect component. If the pure error covariance matrix was assumed to be V–robust w.r.t. the fixed effect model matrix (according to the classical definition), then the total covariance matrix would be structured as (3.4), with $(\boldsymbol{A} - \boldsymbol{B})$ proportional to $\boldsymbol{J}_{16}$; this is the reason why the generalization was preferred (see, for details, the Appendix of MORTARINO and SALMASO (1996)).

4 V–robustness for orthogonal designs with N =80, 112 runs

80 run designs. For $N = 80$ we consider a resolution V design with six main factors and the model matrix obtained from the corresponding design matrix by adding the column **1** and all products between couples of columns. Because of the presence of repeated observations, a V–robust covariance pattern w.r.t. to this matrix would be again a singular one. It might be better to generalize this pattern as showed in the previous section. But the problem is that here we do not have equal repeated blocks of observations and the generalization previously introduced would not be possible. For a detailed discussion of this topic see MORTARINO and SALMASO (1996), sect. 5.

As a consequence, the method used in previous section cannot be applied in this case. Since equal repeated blocks cannot be formed for the presence of the last column in the design matrix, we consider a linear model with a matrix obtained from the original model matrix by eliminating that column (and, consequently products containing it):

$$\widetilde{Z}_{(80\times 16)} = [\,W' \quad W' \quad W' \quad V' \quad V'\,]' \,,$$

where $W = [+1 \quad H_{16} \quad +e \quad P_1]$ and $V = [+1 \quad H_{16} \quad -e \quad P_2]$.

The covariance matrix of the stochastic term of the model is assumed to satisfy these conditions:

$$\sigma_{i(u),j(u)} \;=\; f_A(z'_{i(u)} \odot z'_{j(u)}) \;\; \forall u = 1,2,\ldots,5 \qquad (4.1a)$$

$$\sigma_{i(u),j(w)} \;=\; f_B(z'_{i(u)} \odot z'_{j(w)}) \;\; \forall u,w = 1,2,\ldots,5 \;\; u \neq w. \qquad (4.1b)$$

It can be shown that, in this case, in order to achieve the same result of Theorem 3.1. (and the multivariate extension of Theorem 3.3.), the following requirement about the function f_B of (4.1b) must be introduced: for any $t = 1,2,\ldots,16$,

$$f_B(h'_t, -e'_t) = \sum_{r=1}^{16} \gamma_r f_B(h'_r \odot h'_t, e'_r \odot e'_t),$$

where

$$\gamma_h = \begin{cases} 3/8 & \text{if } h \in \{1,2,3,5,9,16\} \\ -1/8 & \text{if } h \in \{4,6,7,8,10,11,12,13,14,15\} \end{cases}$$

This is equivalent to the following requirement: the covariance between two runs of different groups (which means a run in a block with index in $\mathcal{G}_1 = \{1,2,3\}$ and the other one in a block with index in $\mathcal{G}_2 = \{4,5\}$) must be equal to an algebraic weighted sum of covariances between runs of different blocks belonging to the same group.

112 run designs. These designs have the same structure of the 80 run designs above considered. The condition required for this case is the same found for the 80 run design.

References

[1] ADDELMAN, S. (1961). Irregular fractions of the 2^n factorial experiments. *Technometrics*, **3**, 479–496.

[2] GUSEO, R. and MORTARINO, C. (1997). Multiresponse dependent experiments: robustness of 2^{k-p} fractional factorial designs. Submitted to *Comm. Stat. Th. Meth.*.

[3] KROUSE, D.P. (1994). Patterned matrices and the misspecification robustness of 2^{k-p} designs. *Comm. Stat. Th. Meth.* **23**, 3285–3301.

[4] KOUNIAS, S. and SALMASO, L. (1997a). Experimental designs of resolution *III*, *IV* and *V*. *Proc. of the Annual Meeting of Statisticians of Greece.*

[5] KOUNIAS, S. and SALMASO, L. (1997b). A general characterization of orthogonal designs of different resolution. Rapporto tecnico. Dip. di Scienze Statistiche. Università di Padova.

[6] KOUNIAS, S. and SALMASO, L. (1997c). Orthogonal two level factorial designs of resolution *III*, *IV* and *V*. Submitted to *Jour. of Comb., Inform. and System Sciences.*

[7] MORTARINO, C. (1996). Multiresponse irregular fractions of two-level factorials with dependent runs. In: Forcina A., Marchetti G.M., Hatzinger R. and Galmacci C. eds. *Proc. of the 11th Int. Workshop on Statistical Modelling, Orvieto 15–19 July 1996*, 434–437.

[8] MORTARINO, C. and SALMASO, L. (1996). Extended V–robustness for two level orthogonal resolution five designs. Note di Ricerca 10/96. Dip. di Sc. Statistiche. Università di Udine.

[9] SALMASO, L. (1996). Piani fattoriali frazionati a due livelli di risoluzione V: piani regolari di ordine due. Tesi di Laurea. Facoltà di Statistica. Università di Padova, Padova.

[10] RAO, C.R. (1967). Least squares theory using an estimated dispersion matrix and its application to measurement of signals. In: LeCam, L.M. and Neyman, J. (eds.). *Proc. of the Fifth Berkeley Symp. on Mathematical Statistics and Probability, I.* University of California Press, Berkeley, 355–372.

[11] ZYSKIND, G. (1967). On canonical forms for non–negative covariance matrices and best and simple least squares linear estimators in linear models. *Ann. Math. Stat.*, **38**, 1092–1109.

Computational Algebraic Geometry in Industrial Experimental Design

E. RICCOMAGNO AND H.P. WYNN
University of Warwick

1 Algebraic geometry in the design of experiments

Pistone and Wynn (1996) introduced the idea of Gröbner bases into the design of experiments. This allows the study of "generalised confounding" in non-standard situations, that is for non-standard designs which arise in a number of situations. They were also able to study classical designs such as Plackett-Burman designs from this same view-point.

The starting point is to consider an experimental design as an algebraic variety. Consider for an example the 2^2 experimental design consisting of the four point in two dimensions: $(\pm 1, \pm 1)$. Then one may express this as the solution of $(x_1 - 1)(x_1 + 1) = 0, \quad (x_2 - 1)(x_2 + 1) = 0$. We say that it is the solution given by the intersection of two polynomial ideals.

The main advantages of expressing designs in this way is that the division of polynomials can be used. In general let us take m factors $x_1, \ldots, x_m$ and let $d(x)$ be a polynomial and

$$D = \{x : d(x) = 0\}$$

be a representation of the design as the solution of a system of algebraic equations. The polynomial *ideal* $I(D)$ corresponding to the variety D is the set of *all* polynomials which are zero on the design points D. The basic algebraic theory says that any member of the ideal is finitely generated in that it can be expressed as a polynomial combination of basic generators: $g_1(x), \ldots, g_s(x)$ and

$$D = \bigcap_{i=1}^{s} \{x : g_j(x) = 0\}$$

Now given a general polynomial model $p(x)$ we essentially divide $p(x)$ by the design ideal and return a remainder $r(x)$. This can be written as

$$p(x) = \sum_{j=1}^{s} s(x)_j g_j(x) + r(x)$$

*Kitsos, C.P. and Edler, L., Eds.:Industrial Statistics, Physica-Verlag, Heidelberg, 1997

and forces

$$p(x) = r(x)$$

for all x in the design D. This argument prompts to a clear definition of aliasing or confounding with respect to a design: two models $p_1(x), p_2(x)$ are confounded if and only if they are equal at all $x \in D$. With this argument $p(x)$ and $r(x)$ are confounded. Several points and questions arise naturally form this formulation.

(i) Is there in general a unique set of $g_j(x)$ given the design?

(ii) The remainder in general ought to be of smaller polynomial order than the g_j.

(iii) It would be pleasing if the remainder were unique, under some conditions so that it is in some sense a unique representative lower order for the class of confounded polynomials. These questions are resolved with the use of Gröbner bases. A basic readign is Cox, Little and O'Shea (1992).

2 The background algebra

Let $\mathbb{Q}[x_1, \ldots, x_m]$ be the set of all polynomials in the indeterminates $x_1, \ldots, x_m$ with rational coefficients. Polynomials of the form $x_1^{\alpha_1} \cdots x_m^{\alpha_m}$ are called monomials and play an important role in the theory. Let $f_1, \ldots, f_k$ be polynomials in $\mathbb{Q}[x_1, \ldots, x_m]$. The polynomial ideal generated by $G = \{f_1, \ldots, f_k\}$ is the set of all polynomial combinations and we write it as

$$I = < f_1, \ldots, f_k > = \left\{ \sum_{i=1}^{k} g_i f_i : \quad g_i \in \mathbb{Q}[x_1, \ldots, x_m] \right\}. \tag{1}$$

The set G is called a basis of I.

Let $D = \{(a_{i_1}, \ldots, a_{i_m}) : i = 1, \ldots, N\}$ be a design with N points with rational coordinates in m factors. Let us consider a system of equations whose solutions are exactly the design points. There are different ways to determine this set, see Pistone and Wynn (1996). We proceed as follows. The design ideal for the one point design variety of coordinates $(a_1, \ldots, a_m)$ is

$$< (x_1 - a_1), \cdots, (x_m - a_m) >$$

The variety for a design with more points is the union of the single point varieties and the ideal is the intersection of the corresponding one-point ideals. All this says is that the set of polynomials through all the design points is the intersection of the larger sets though each individual point. We call $I(D)$ the design ideal generated by D. The procedure for intersecting ideals is implemented e.g. in CoCoA.

A monomial ordering is a total order, $\succ$ on the set of all monomials satisfying the following condition

$$x_1^{\alpha_1} \ldots x_m^{\alpha_m} \succ x_1^{\beta_1} \ldots x_m^{\beta_m} \text{ implies } x_1^{\alpha_1+\gamma_1} \ldots x_m^{\alpha_m+\gamma_m} \succ x_1^{\beta_1+\gamma_1} \ldots x_m^{\beta_m+\gamma_m}$$

for all $(\alpha_1, \ldots, \alpha_m)$, $(\beta_1, \ldots, \beta_m)$ and $(\gamma_1, \ldots, \gamma_m) \in \mathbb{Z}_+^m$.

For example given an initial ordering on the indeterminates $x_1 \succ \ldots \succ x_m$ then for $\succ$=**DegRevLex** $x_1^{\alpha_1} \ldots x_m^{\alpha_m} \succ x_1^{\beta_1} \ldots x_m^{\beta_m}$ if its total degree (sum of the exponents) is greater or its total degree is equal and it precedes $x_1^{\beta_1} \ldots x_m^{\beta_m}$ in the inverse lexicographic order, which is when **DegRevLex** finds the same total degree it looks at the smallest variable in the initial ordering and favours the smaller power.

The remainder term $r(x)$ is generic in that as $p(x)$ ranges over all polynomials in m variables we obtain the vector space spanned by the monomial terms making up a typical $r(x)$. Algebraically this vector space is the *quotient ideal* and the mo nomial terms generate the *coordinate ring.* The monomials in $r(x)$ for a given monomial ordering are unique and are found as follows. First let L be the set of leading terms of the G-basis $\{g_j\}$ of the design ideal $I(D)$. Then the monomial terms making up $r(x)$ are *the set of monomials not divisible by any term in L.* The terms in $r(x)$ satisfy a divisibility condition: if a particular monomial is included then any monomial which divides that monomial is also included. This is à natural condition for modelling since it is common to include in a model lower order terms if higher order terms are include (for example in backwards stepwise regression).

3 Fans and Computer experiments

Computer experiments is the expression which refers to the fitting of statistical models to computer code (simulator) for use in fast sensitivity analysis, root-finding, inverse engineering and optimisation. A typical application is to computer-aided design. Thus if

$$y = f(x)$$

is some complex function known via a computer program one may observe $y_i = f(x_i)$ at sites $x_1, \ldots, x_N$ and fit an emulator $\hat{y}_i = \hat{f}(x_i)$. Since typically the observations are made without experimental error it is natural to ask for the interpolation property that the emulator agrees with the simulator at the design sites:

$$\hat{f}(x^i) = f(x^i), \quad (i = 1, \ldots, N)$$

Since this exact interpolation property holds for the interpolator $r(x)$ in the above analysis it is natural to consider them as candidates for emulation.

When there are competing $r(x)$ for the same design and data (x^i, y^i), $(i = 1, \ldots, n)$ and since there are no errors of observation in the normal sense one must resort to some other methods of model choice. In other work the authors consider Bayesian methods of model choice. Here we consider a criterion based on smoothness.

Consider a very simple example in one dimension to clarify the issues. Consider the design $\{-1, 0, 1\}$ and two alternative interpolators $p_1(x), p_2(x)$. Let

$p_1(x)$ be based on the $r(x)$ terms $\{1, x, x^2\}$ and let $p_2(x)$ be the model with the linear term replaced by a cubic term $\{1, x^2, x^3\}$. Suppose the data are respectively $\{y_1, y_2, y_3\}$. Take as the measure of smoothness:

$$e_j = \int_{-1}^{1} p_j''(x)^2\, dx, \quad (j = 1, 2)$$

Then a simple computation gives that

$$e_2 - e_1 = 6(y_1 - y_3)^2 \geq 0$$

showing that the lower order interpolator p_1 is smoother.

The most natural generalisation of this idea is to choose between models on the basis of the integrate square Gaussian curvature:

$$\int_{\mathcal{X}} \kappa^2 dx$$

In Caboara, Pistone, Riccomagno and Wynn (1997) the idea of a fan of an experimental design is presented. It is the set of estimable models $r(x)$ subject to the divisibility condition mentioned in Section 2. Each $r(x)$ is called a leaf and different leaves are obtained by changing the monomial ordering. A maximal fan design is one for which every possible leaf can be estimated. Such designs tend to be space filling and look very similar to the "latin hypercube" designs used in computer experiments.

Now if every leaf is a candidate for an interpolator or emulator of the underlying model $y = f(x)$ then this emphasises the importance of finding a method such as the use of curvature for choosing between emulators. Suppose for example the model is almost "flat" in all except one dimension; that is there is considerable variation in the x_1 direction but not in the $x_2, \ldots, x_m$ directions. A classical experimental design say with 2 levels would not allow this to be found. With 2 levels only the linear variation in the x_1 direction could be fitted. With a maximal fan design the model with one dimensional polynomial terms $\{1, x_1, \ldots, x_1^{N-1}\}$ could potentially be fitted. However without a method for selecting this model over all other potential models $r(x_1, \ldots, x_m)$ it is difficult to see how this could be carried out.

4 Models as varieties

Over and above the use in experimental design considering models as varieties has considerable advantages in the understanding of modelling issues. Consider a polynomial model (or rational polynomial) for a mean response

$$\eta = p(x, \theta)$$

where x and θ are, possibility multivariate, independent variables and parameters respectively. This is conceptually identical to the models above with error except, to emphasise the point, we replace Y by η. An additional aspect is that $p(x, \theta)$ may be nonlinear (though still rational polynomial). More general still is to incorporate the response term and write:

$$F(\eta, x, \theta) = 0$$

where F is polynomial. When the original model is rational we can multiply out and prevent denominators being zero by additional restrictions in an obvious way. Then both combined experimental design and data point $\{x_i, y_i\}$ can be expresed as an ideal and incorporated in the same way as the design point were above.

Note that in all this the division of variables into different categories: designed (controlled), observed, unobserved, arise from human cognitive intervention. The mathematics really only acknowledges the terms: numerical/known and unknown. Thus the purpose is to express the unknown in terms of the known. This is very close to being a problem in pure computer science except that distinctions are made in the case of random variables versus purposive control ones. Such variables may be dealt with differently in the analysis via likelihood approach.

4.1 Physical models

The algebraic geometry approaches described above for modelling allow the study of models which come from different areas sometimes referred to as mechanistic models. A hard problem is how to combine the empirical models of the statistician with the models arising from the underlying science or technology from the point of view of experimental design. The present approach gives some understanding of the structural relationships between data and unobservables similar to state-space models in system theory.

Consider two models one the model $F = 0$ above and the other

$$\tau(\eta, x, \phi)$$

given by, say, physical considerations or some restriction on the first model.

The independent and response variables x and η are the same but the parameters θ and ϕ are possibly different. For simplicity we return to our old form of $F = 0$ and write $y - p(x)$ and concentrate in our expalntion on the x variables. Using the *normalf* procedure on Maple we can find the remainder $r(x)$ of one mo del with respect to the other. The procedure is first to re-express the second model in G-basis form $\{x : p(x) = 0\} \Leftrightarrow \bigcap_j \{x : g_j(x) = 0\}$. Then express the first model as

$$p(x) = \sum_j s_j(x) g_j(x) + r(x)$$

Then on the model $\pi(x) = 0$ we have $p(x) = r(x)$. Normalf gives $r(x)$.

Suppose now that the input variable x has variation but that $\pi(x) = 0$ is assumed to hold exactly, without variation. Then the variation in $p(x)$ is identical to that in $r(x)$ together with the added restriction imposed by $\pi(x) = 0$. It is possible to use conditional random variables to describe this but here we take a simple local approach. Consider the total derivative of $r(x)$:

$$dr = \sum_{i=1}^{m} \frac{\partial r}{\partial x_i} dx_i$$

which expresses the variation in $r(x)$ for small changes in the x_i. We can do the same for the g_j having in mind now that the dg_j are zero:

$$\sum_{i=1}^{s} \frac{\partial g_j}{\partial x_i} dx_i = 0 \quad (j = 1, \ldots, s)$$

This imposes additional restrictions on the dx_i and hence on the variability in dr.

All this can be extended to the more general implicitly defined model F. It can be used to place extra assumptions on the output η or parameters θ. In the latter case it generalises the familiar generalised liner model situation in which linear constraint on parameters are induced through to the model. Restrictions on output are familiar in tolerancing problems and the effect on the model and input space for x can be thought of inverse engineering. In fact many of the applications in engineering, for example to robotics, are of this kind.

An important source of assumption arise from group invariance restrictions on models. In this case the π may represent the maximal invariants for some physical model. For example for rotation in R^2 we have $x_1^2 + x_2^2 = c$ where c is a constant.

References

[1] Pistone, G., Wynn, H. (1996). Generalised confounding with Gröbner bases. *Biometrika*, **83**(3):653–666.

[2] Caboara, M. and Pistone, G. and Riccomagno, E. and Wynn, H.P. (In progress). Generalised confounding II: the fan of an experimental design.

[3] Cox, D., Little, J., O'Shea, D. (1992). *Ideal, Varieties, and Algorithms.* Springer-Verlag, New York.

Part IV

Pharmaceutical Statistics

New Statistical Methods for Analyzing Mutagenicity Assays: Real Data Problems in Biopharmaceutical Drug Development

L. A. HOTHORN

University of Hannover, Herrenhauser Str. 2.,
D-30419 Hannover, Germany

Key words:
mutagenicity, trend test, order restriction, proof on safety

Abstract:
Both adequate formulation of the hypotheses (difference vs. equivalence) and real data problems of a mutagenicity assay (a-priori unknown distribution, tied values, small sample sizes, non-monotonic dose-response relationship, etc.) are discussed.

1. Introduction

The approval of a new drug depends on both proof on efficacy in clinical trials and proof on safety in toxicological studies. One of the frequently used and highly standardized studies represents the Ames mutagenicity assay. However, the statistical principles discussed below can be used for other mutagenicity assay's, e.g. micronucleus, dominate lethal and chromosome aberration assay, in a similar way. The objective of the Ames-assay is the determination of a potential gene-mutagenic effect of a substance after single administration of several doses on

*Kitsos, C.P. and Edler, L., Eds.:Industrial Statistics, Physica-Verlag, Heidelberg, 1997

histidine-free agar plates with specially modified *Salmonella typhimurium* bacteria strains. The number of grown visible colonies (so called revertants) of the dose groups are compared with those of the negative and positive control. According to international guidelines several strains are used with different sensitivities for the different DNA mutagenic mechanisms (e.g. base change mutation using TA 100 strain). The design is a completely randomized one-way layout [C-, D_1, ..., D_k, C+], where the number of doses k is between 3 and 6. The positive control C+ represents a substance with known mutagenic effect, but moderate dosed. Usually low sample sizes (number of plates) are used, frequently unbalanced because of increased sample size in the negative control. Since the false negative rate increases dramatically using too low doses (and the doses are arbitrary in this *in-vitro* assay in relation the human doses), the doses have to be selected in such a way that a non-monotonic dose-response relationship occurs due to the interaction of mutagenic and cell-toxic effects (so called muta-tox problem). Therefore, an „umbrella" shape of the dose-response is possible. At least two independent experiments should be used. In Tab. 1, an example with raw data are given.

Table 1: Teaching example: raw data of an Ames assay on TA 100

j	Dose	Revertants	$\overline{x}_j$
0	0	80, 83, 78, 88, 78	81.4
1	0.01	93, 88, 78	86.0
2	0.02	97, 103, 99	98.7
3	0.04	104, 103, 95	100.7
4	0.08	95, 97, 99	96.7
C+		294, 333, 333	320.0

This assay was established already in 1975 and publications on statistical analysis starts in 1978 (see Edler, 1992 for an overview). However, until now some concerns in routine analysis exist. A statistical method for such a routine industrial application differs from a methods used in basic research. It should be:

- a statistical approach (taken variability into account), not only a fixed point decision rule like the two-fold rule
- robust against violations of the underlying assumptions,

- based on recommendations from the international harmonization documents
- widely accepted (not an extreme new approach) and as simple (and cheap) as possible,
- implemented in validated software according to GLP and ISO9000,
- used as less as possible a-priori model assumptions (because rather different substances with different mechanisms will be investigated routinely),
- taken into account, that holding the null-hypothesis is the most frequent outcome,
- easy to understand by toxicologists and physicians (the real decision makers),
- given the correct answer with defined false negative (in the case of holding null-hypothesis) and false positive rates (in the case of rejection of null-hypothesis).

Recently, recommendation for analyzing these assay were published by Hauschke et al. (1997) taken into account the above principles.

2. Proof and Hazard Versus Proof on Safety

Usually, the traditional null hypothesis of no difference in the effect between the doses and the negative control is tested. Failing to reject the null hypothesis usually leads to the conclusion of evidence in favor of safety. However, frequentist's testing theory based on the falsification principle: the proof of an effect is only possible to demonstrate the low probability of its contradiction. Therefore, the major drawback of frequently used indirect approach is that the probability of erroneously concluding hazard, i.e. producer's risk, is controlled directly only. However, the primary concern of safety assessment is the control of the consumer's risk, that is controlling the probability of erroneously concluding safety. Thus, the adequate test problem should be formulated by reversing the null hypothesis and the alternative and incorporating an a-priori or a-posteriori defined threshold, i.e. equivalence testing in the k-sample problem. This direct approach was demonstrated for the *k*-sample many-to-one problem with respect to non-monotonic and monotonic response relationships by Neuhäuser and Hothorn (1997) and Hauschke and

Hothorn (1997). This approach seems favorable in the near future, because for the single endpoint „number of revertants" a threshold can be defined a-priori based on either the two-fold rule or on reference values using the historical controls. However, this approach is new and until now neither used in routine analysis nor accepted by authorities. Therefore, in chapter 3 modification for the traditional testing procedure on difference (proof on hazard) is described. However, one objective is obvious: selected an as most as powerful approach (minimum type II error = consumer's risk).

3. Real Data Problems in the Testing Procedure on Difference

The primary objective is a global test with an ordered alternative and the result: „trend or no trend", i.e. „potential Ames positive or negative". Local answers, i.e. NOED (no observed effect dose), or estimation the magnitude are discussed controversially in the literature until now. The trend test should be robust against downturns at high doses, where the umbrella point is a-priori unknown. Several approaches were published, e.g. protected trend tests (Simpson and Margolin, 1986); umbrella tests (Chen, 1992) or partial order restriction (Hothorn, 1997). However, looking on a long experience in routine analysis, at least 10% of the assay's reveal a non-monotonicity. Because this effect is unknown a-priori, all assay's have to be analyzed by trend test robust against non-monotonicity. Therefore, the question arises, is a loss of power for 90% or more of the assay's acceptable? To look on the data and decide the type of trend test represents „hunting for significance" from a theoretical viewpoint. However, this empirical approach never decrease power, and hence should be accepted until a better approach is available. One such possibility seems to be so-called multiple contrast tests consisting of one contrast test sensitive for concave and another for convex shapes simultaneously. First results (Neuhäuser and Hothorn, 1997) indicate that this represents an „overall powerful" test. More detailed investigations for the specified condition are needed, however.

Numerous publications try the characterize the distribution of the endpoint number of revertants (or others, e.g. number of micronuclei). However, based on the very small sample sizes of the concurrent assay, i.e. $n_j=3,...,5$ and the routine character, a powerful test to characterize the distribution seems hopeless. A possible way is the used of historical control data, because mutagenicity assay's are highly standardized within a laboratory. However, only the distribution for negative control can be estimated, never for the dose groups. Therefore, at general non-parametric tests should be recommended, although the alternative hypothesis „stochastic order“ seems not to be the experimental question and is hard to understand by toxicologists. Because of the very small number of replications („at least three“ recommended by the guideline), the possible unbalancedness and the possible ties (depending on the magnitude of the number of revertants, depending on the used strains), exact non-parametric tests are recommended. Now, efficient algorithms are available to calculate selected contrast tests with up to 5 groups and replications up to 6 by total enumeration (SAS, STATXACT). Moreover, for higher dimensions and/or sample sizes, simulation-based algorithms are available, too (Berry, 1995).

A critical point represents a possible increase of variance with increasing effects. Although this effect occurs frequently in real data, and parametric tests based on MSE and (both pairwise and k-ranking) non-parametric tests loose power dramatically, this problem is frequently ignored in routine analysis. Until now no „overall-robust“ approach seems available. One possibility seems to be a modification of Behrens-Fisher rank transformation approach according to Brunner and Puri (1996). Another possibility to treat both the distribution and variance problem is the assume a mixing distribution of „responder and non-responder (non-responder behavior like C-)“ according to Hothorn (1994).

The predicative value of the Ames assay is limited by its disposition to an increased false-positive rate (e.g. due to small variance of the tied data and biological arguments). Moreover, a formulation of the classical hypotheses with threshold zero seems not apprppriate in biopharmaceutical research at general. Therefore, a relevance-shifted hypothesis seems favorable: H_0: $\mu_D - \mu_C \leq \delta$ vs. H_A: $\mu_D - \mu_C > \delta$.

However, the a-priori definition of the threshold δ is controversial; e.g. the two-fold rule can be used.

An open point is the reproducibility of the assay based on at least two assay's. To my knowledge, nothing is known on combination of independent studies under the aspect of the safety approach. It seems hard to accept to conclude „Ames negative" according to Fisher's product criterion, if the p-values of both studies are $p_1=0.009$ and $p_2=0.3$, for instance.

The inclusion of the positive control into the statistical analysis using an ANOVA approach was warned by Kirkland (1989). A simple stepwise testing approach was published by Hothorn (1995) for this design. The hypotheses are ordered, starting with the test on sensitivity by comparison C- vs. C+ followed by tests on substance effects C- vs. D_j. Both tests can be done at level α, but if the hypothesis on sensitivity is not rejected, the procedures stops and no further test on substance effect is allowed (closure principle).

4. The Example

Here, the analysis of the example in Table 1 is described. A-priori no information on the distribution of the number of revertants in these experimental groups is known. The sample sizes are small, even simple Box-plots give no information on the distribution. Therefore, non-parametric permutative tests are used. First, comparing C- and C+ as a test on sensitivity of the concurrent assay is performed by a one-sided permutative U-test. The p-value is $p_{sens.}= 0.018$ and therefore the global trend test can be performed at level α. The usual way is the analysis by any non-parametric trend test, e.g. Helmert contrast test. The p-value is $p_{Helmert}=0.21$. However, Helmert contrast is not powerful for concave and „umbrella" shapes. Therefore, the multiple contrast test max{Helmert contrast; reverse Helmert contrast} as a permutative version was used and a p-value=0.004 results. Moreover, a relevance shifted test modification is used with an a-priori defined threshold of $\delta=8$ and a p-value of 0.10 results. The outcome of the relevance shifted test („potential Ames negative") is supported by the two-sided 95% tolerance limits for

95% of the future population for this strain [92; 146]. No single value is larger than the upper limit.

5. Conclusions

The routine analysis of Ames mutagenicity studies needs a robust and relevance-shifted test. Here a simple approach, robust against the a-priori unknown distribution and unknown shape of the dose-response relationship is described. A modification for heterogeneity of variance is still needed, however.

References

Berry, J.J. (1995) A simulation-based approach to some nonparametric statistics problems. *Observation* 5, 19-26.

Brunner, E. and Puri, M.L. (1996) Nonparametric methods in design and analysis of experiments. In: Handbook of Statistics (Gosh, S; Rao, C.R. eds.) Vol. 13. Elsevier Sci. B.V. Amsterdam

Chen, Y.I. (1992) Robust umbrella tests for a generalized Behrens-Fisher problem *Biometrical J.* 34, 347- 357

Edler L. (1992): Statistical methods for short-term tests in genetic toxicology. The first fifteen years. *Mutation Research* 277, 11-33.

Hauschke, D. and Hothorn, L. (1997) Safety assessment in toxicology studies: proof of safety versus proof of hazard. In: Chow, S.-C. and Liu, J.-P. (eds.): *Design and analysis of animal studies in pharmaceutical development*. Chapter 6. Marcel Dekker, New York (in press)

Hauschke, D.; Hayashi, M.; Lin, K.K.; Lovell, D.P.; Robinson, W.D.; Yoshimura, I. (1997): Recommendations for biostatistics of mutagenicity studies. *Drug Information J.* 31, 232-326.

Hothorn, L. (1994) Biostatistical analysis of the micronucleus mutagenicity assay based on the assumption of a mixing distribution. *Environ. Health Perspect.* 102 Suppl. 1, 33-38.

Hothorn, L. A. (1995) Biostatistical analysis of the experimental design 'control versus k treatments' including a positive control group. In: Vollmar, J. (ed.) *Biometrie in der chemisch-pharmazeutischen Industrie*, Volume 6, Fischer Verlag Stuttgart, 19-26.

Hothorn, L.A.(1997) Modifications of the closure principle for analyzing toxicological studies. *Drug Information J.* 31, 403-412.

Hothorn, L.A. and Hauschke, D. (1997) Statistical Testing Principles in Randomized Toxicological Studies In: Chow, S.-C. and Liu, J.-P. (eds.): *Design and analysis of animal studies in pharmaceutical development.* Chapter 4. Marcel Dekker, New York (in press)

Kirkland, D. J. (Ed) (1989): *Statistical Evaluation of Mutagenicity Test Data.* Cambridge University Press, Cambridge.

Neuhäuser, M. and Hothorn, L.A. (1997) The control of consumer's risk in the Ames mutagenicity assay. *Drug Information J.* 31, 363-368.

Simpson, D. G. und Margolin, B. H. (1986): Recursive nonparametric testing for dose-response relationship subject to downturn at high doses. *Biometrika* 73, 589-596.

Application of Statistical Selection Procedures in Biotechnology

Ute RÖMISCH[1] and S. GARGOVA[2]

[1] Technische Universität Berlin, FB Lebensmittelwissenschaft und Biotechnologie, FG Informatik, Invalidenstr. 42, D-10115 Berlin, Germany
[2] Higher Institute of Food and Flavour Industry, Plovdiv Dep. of Biotechnology, Bulv. Maritza 26, B- 4000 Plovdiv, Bulgaria

Key words:
biotechnology, enzyme activity, selection, subset, indifference zone procedure

Abstract:
A short methodical summary about the two classes of statistical selection procedures is given, i.e. the indifference zone (and d-correct) procedures and the subset procedures. The biotechnological problem consisted in selecting the „best" or at least a „good" mutant with high enzyme activity from a set of eight mutants of the species *Aspergillus niger* with a large probability. Depending on suppositions about the variances of the enzyme activities, different selection rules are applied. Starting with the subset procedure of Gupta (1985) for the case of equal variances, the number of mutants is reduced. The following d-correct procedure of Bechhofer, Dunnett and Sobel (1954) calculates the necessary sample size n. Then the mutant whose sample has the largest mean will be selected as a „good" one with a given precision of d and a probability of correct selection of (1-ß). As long as the variances of the enzyme activities are assumed to be different, the selection procedure of Dudewicz and Dalal (1975) must be used.

1. Introduction

The enzyme Glucoamylase is very important in food industry. It is used for producing glucose, ethanol, and beer. Eight mutants of the species *Aspergillus niger* have been evaluated and tested with respect to their production of the enzyme Glucoamylase and proved to be suitable.

For the mutants M_i (i=1,...8) the enzyme activities Y_i (i=1,...,8) are observed. It can be supposed that the enzyme activities Y_i are normally distributed with unknown population means μ_i and variances σ_i^2 (i=1,...,8). The practical problem now consists in selecting the „best" or at least a „good" mutant from the set of mutants on the basis of eight samples (Y_{i1} ,..., Y_{in_i}) of size n_i (i=1,...,8). The mutant M_i is denoted as „better" than the mutant M_j (i≠j, i,j=1,...,8) if its population mean μ_i is larger than μ_j .

There are two possible formulations of the selection problem:

1. A decision about the „best" or at least a „good" mutant is required. Such a decision must be connected with a given probability of its correctness. The experimenter must define a difference d between the largest population mean and the other means, which is of practical relevance for him. Then the „best" mutant can be selected with a large probability if it is in fact essentially better than the other mutants. In case the „best" mutant is only a little bit better than the „second best", a very large sample size would be necessary for the selection of the „best" mutant. Hence, it is sufficient to select a mutant whose population mean deviates from the largest population mean by less than the given relevant difference d. This mutant is then called a „good" one. It may be the „best" but it may be also only one of the best. The class of procedures for solving this problem are the indifference zone and the class of d-correct procedures.
2. The aim is to reduce the whole set of mutants so that a subset of mutants containing the „best" mutant with a high probability will be selected. The class of procedures for handling this problems are the subset procedures.

2. General Remarks on Selection Procedures

There are two groups of selection procedures: the indifference zone (and d-correct) procedures and the subset procedures. Let us consider normally distributed populations $Y_i \sim N(\mu_j , \sigma_i^2)$ (i=1,...,k) and assume that the population means are ranked: $\mu_{(1)} \leq \mu_{(2)} \leq ... \leq \mu_{(k)}$; but it is not known which population is associated with the $\mu_{(i)}$ (i=1,...,k). The restriction on normal distributions is possible, because selection rules based on sample means are robust against deviations from the normal distribution. Otherwise there are also nonparametric procedures that do not suppose normal distributions of the populations (Sobel,1967, Hsu, 1980,1981).

The solution of a selection problem by an ***indifference zone procedure*** was developed by Bechhofer (1954) and is defined as follows: Select a population from the k given populations in such a way that the selected population is the „best" one (with the largest population mean) with a probability of at least (1-β) as long as the difference $\mu_{(k)} - \mu_{(k-1)} \geq d$ for a given positive real number d and a given probability (1-β). This condition must be fulfilled for a correct selection of the best population, but it is unknown in practice.

The ***d-correct procedures*** do not use this assumption. Here a „good" population Y_i (i=1,...,k) is selected with a probability of at least (1-β) if $\mu_{(k)} - d < \mu_i \leq \mu_{(k)}$ is valid for μ_i. This means that a good population with a large, but not necessarily the largest population mean will be selected. It is only the „best" one if the condition of the indifference zone procedures is fulfilled. This is the reason why all d-correct procedures contain indifference zone procedures, but the inversion does not hold. For some indifference zone procedures, Chiu (1974) has shown that they too are d-correct procedures. Indifference zone and d-correct procedures enable the calculation of the sizes of the samples to be taken from the k populations. The sample size depends on the practically relevant difference d, the probability (1-ß), the number k of populations and the variances σ_i^2. Then a selection rule is used for selecting the „best" or a „good" population.

Depending on the assumptions about the variances of the k populations, the following selection procedures are recommended:

1. The k populations have the same known variance.
   - Application of the selection procedure of Bechhofer (1954)
2. The k populations have unknown but equal variances.
   - Application of the selection procedure of Bechhofer, Dunnett and Sobel (1954)
   - Application of the selection procedure of Gupta and Kim (1984)
3. The k populations have unknown and unequal variances.
   - Application of the selection procedure of Dudewicz and Dalal (1975)

Cases 2 and 3 are of particular practical interest. All procedures mentioned above, except for the procedure of Gupta and Kim, are indifference zone procedures as well as d-correct procedures.

Subset procedures select a subset from the k populations containing the population with the largest population mean with a probability of at least (1-β). The number of populations contained in the subset should be as small as possible. In general, the calculation of the sample sizes is not a part of this procedures. The advantage of these subset procedures is that they can be used for samples of arbitrary size. The number of populations contained in the subset depends on the probability (1-β), the sizes of the samples, the number k of populations and the variances σ_i^2. The subset selection procedure of Gupta (1985) is used for selecting a subset of normal populations with unknown but equal variances.

There are not only pure indifference zone procedures and subset procedures. For instance, it could be useful to apply, in a first step, a subset procedure for selecting a subset from a large set of populations. In a second step, it would be possible then to apply an indifference zone or d-correct procedure

for selecting the best or a good population to the populations of the subset only. In this way, the total sample size can be reduced.
The following selection procedures are used for solving the biotechnological problem.

2.1. Selection procedure of BECHHOFER, DUNNETT and SOBEL

This d-correct procedure is used for selecting a good population from k normal populations with unknown but equal variances. A relevant difference d to the largest population mean $\mu_{(k)}$ is given, all differences less than d are not of practical interest. The aim is to select a population whose population mean is $\mu_{(k)}$ or larger than $\mu_{(k)} - d$ with a probability of (1-β). At first the procedure calculates the necessary sample sizes. Then, on the basis of the sample means, the population which belongs to the largest sample mean will be selected as a good one. The procedure consists of two steps.

<u>Step 1:</u>
We take samples of the same size n_0 with $6 \le n_0 \le 30$ from the k populations and calculate the means $\bar{y}_1(n_0),...,\bar{y}_k(n_0)$,
the variances $s_1^2(n_0),...,s_k^2(n_0)$
and the common estimated variance $s^2 = \frac{1}{k}\sum_{i=1}^{k} s_i^2$.

Then the necessary sample size is calculated as:

$$n = \max\{n_0\,;\, g\}, \quad \text{where } g = \text{int}\,(2\left(\frac{s}{d}t\right)^2 + 1)$$

with $t = t(k\text{-}1;\, f;\, \rho\,; 1\text{-}ß)$, $f = k(n_0\text{-}1)$, $\rho = 0.5$.
The value t is the (1-ß)-quantile of the (k-1)-dimensional t-distribution with the degree of freedom f and the correlation coefficient $\rho = 0.5$. Tables of the (k-1)-dimensional t-distribution can be found in table 3a of Horn et al. (1995) or table 1 of the procedure 1/51/1012 of Rasch et al. (1996). The function „int" defines the rounding to the next lower integer.

If $n = n_0$ → the population whose sample (on the basis of n_0 observations) has the largest mean will be selected as a „good" one with a probability of (1-ß) and the procedure stops.

If $n = g$ → a second step is necessary.

<u>Step 2:</u>
Another samples of size $g - n_0$ must be taken from each of the k populations and the population whose sample (on the basis of n values) has the largest mean will be selected as a „good" or even the „best" one with a probability of (1-ß).

2.2. Selection procedure of GUPTA and KIM
This procedure is a combination of a subset procedure and an indifference zone procedure. It is used for selecting the „best“ population from a set of normal populations with unknown but equal variances. If the difference of the largest and the second largest population mean is at least larger than the given relevant difference d, the aim is to select this best population with a probability of (1-ß). This supposition seems to be fulfilled if the difference of the largest and the second largest sample mean is sufficient large.

Step 1: (Subset procedure)
We take samples as in the Bechhofer procedure. Now we select those populations M_i, for which:

$$\bar{y}_i(n_0) \geq \max_{1 \leq j \leq k} \bar{y}_j(n_0) - \max\{0 ; t s \sqrt{2/n_0} - d\} \qquad i=1,\ldots,k$$

with $t = t(k\text{-}1; f; \rho; \sqrt{1-ß})$, defined as above.
If only one population is selected, then it is the "best" one and the procedure stops. If more populations are selected, a second step follows.

Step 2: (Indifference zone procedure)
We calculate the necessary sample size n as follows:

$$n = \max\{n_0 ; g\}, \text{ where } g = \text{int}\left(2\left(\frac{s}{d}\right)^2 + 1\right)$$

If $n = n_0$ → the population whose sample has the largest mean will be selected as the "best" one with a probability of (1-ß).

If $n = g$ → as in BECHHOFER et al. above $g - n_0$ values from each population will be added.

2.3. Selection procedure of DUDEWICZ and DALAL
This procedure is used for selecting a good population from k normal populations with unknown and unequal variances. The aim is to select a population whose population mean is $\mu_{(k)}$ or larger than $\mu_{(k)} - d$ with a probability of (1-β).
The procedure includes the calculation of the necessary sample sizes in dependence on the different estimated variances. Then the sample means based on these sizes are calculated and the population which belongs to the largest sample mean will be selected as a good one.

Step 1:
We take samples of the same size n_0 with $6 \leq n_0 \leq 30$ from the k populations and calculate the variances $s_1^2(n_0),\ldots,s_k^2(n_0)$.
Now the necessary sample sizes are calculated:

$$n_1 = \text{int}((h/d)^2 s_1^2 + 1) \quad ,\ldots, \quad n_k = \text{int}((h/d)^2 s_k^2 + 1).$$

Tables with h- values depending on k, n_o and β can be found in Horn (1995) and Dudewicz and Dalal (1975).

If $n_i \leq n_o$ → the means are calculated on the basis of the existing n_o sample values

If $n_i > n_o$ → a second step is necessary.

Step 2:

Another samples of size $n_i - n_o$ must be taken from the populations M_i for calculating the sample means. The population whose sample has the greatest mean now will be selected as a „good" or the „best" one with a probability of (1-ß).

2.4 Selection procedure of GUPTA

The idea of Gupta (1985) is to construct a subset of a low size in such a way that it contains the population with the largest population mean with a probability of (1-β). If the sizes of the samples are low, the subset consists of a large number of populations and conversely. This procedure can be used for normal populations with unknown, but equal variances.

We take samples from the k populations and calculate

the means $\bar{y}_1, \ldots, \bar{y}_k$,

the variances $s_1^2, \ldots, s_k^2$

and the common estimated variance $s^2 = \frac{1}{k}\sum_{i=1}^{k} s_i^2$.

With a probability of (1-β) the subset then contains the populations for which:

$$\bar{y}_i \geq \max_{1 \leq j \leq k} \bar{y}_j - ts\sqrt{2/n} \quad ,$$

with $t = t(k-1; f; \rho; 1-\beta)$.

3. Biotechnology Example

The above selection procedures will be illustrated by their application to the problem of selecting the mutant with the highest enzyme activity and an acceptable stability from a set of eight different mutants of the species *Aspergillus niger*. Detailed experimental methods of cultivation, the syntheses of Glucoamylase, and the determination of its activity are described in Gargova et al (1984). A mutant whose population mean has a difference from the largest mean of less than the relevant difference d = 4 is called a „good" one. Tab. 1 contains the enzyme activities [u/l] of the eight mutants, the sample means and the variances. Fig. 1 provides a survey on the 95% confidence intervals of the means and Fig. 2 contains the Box- and Whisker-Plots with the minimal and maximal values (($y_{(1)}$) and ($y_{(n)}$)), the medians $\tilde{y}_{0.5}$, the quartiles $\tilde{y}_{0.25}$ and $\tilde{y}_{0.75}$, the interquartile ranges I and the ranges R .

Tab. 1. Experimental results

No.	Enzyme activities of the mutants [u/l]							
j	M1	M2	M3	M4	M5	M6	M7	M8
1	53.93	46.64	53.93	41.65	46.64	36.23	51.64	37.90
2	54.35	60.39	53.93	57.06	54.35	46.64	51.64	49.98
3	62.05	59.19	53.93	53.93	53.93	32.16	53.93	53.93
4	66.64	62.46	54.15	58.31	26.66	41.65	54.15	50.00
5	54.15	62.46	58.31	54.15	41.65	37.46	54.15	62.46
6	54.15	45.82	54.15	37.46	58.31	33.32	54.15	62.46
7	26.66	45.82	37.46	26.66	37.46	54.15	62.46	54.15
8	33.32	58.31	45.82	45.82	54.15	26.66	66.64	70.80
$\bar{y}_i$	50.66	55.14	51.46	46.88	46.64	38.53	56.09	55.21
s_i^2	187.20	58.16	43.93	123.27	116.22	76.36	29.63	100.72

The variances of the mutants M1, M4, M5 and M8 seem to be larger than the variances of the other mutants (Tab.1). Although we could suppose equality of the variances as a result of the Bartlett-test as well as the Cochran-test
($\alpha = 0.05$), we decided to use and to compare selection procedures for equal and unequal variances because the sample sizes were with ni = 8 rather small.
It can be concluded that the sample means and the medians of the mutants M2, M7 and M8 are larger than the corresponding parameters of the other mutants. The mean and the median of the mutant M6 are smaller than of the others.
Under the supposition of homogeneity of the variances it is reasonable at first to use the subset procedure of Gupta to reduce the number of populations. Then the d-correct procedure of Bechhofer et al. is applied to select a good population from the subset of populations. The application of the selection procedure of Gupta and Kim will be shown next. It is uncertain whether the difference between the „best" and the „second best" mutant is larger than a relevant difference d, because the sample means of M2, M7, and M8 do not essentially differ and therefore the necessary condition for an indifference zone approach probable is not fulfilled.
For the case of unequal population variances the d-correct procedure of Dudewicz and Dalal is used.

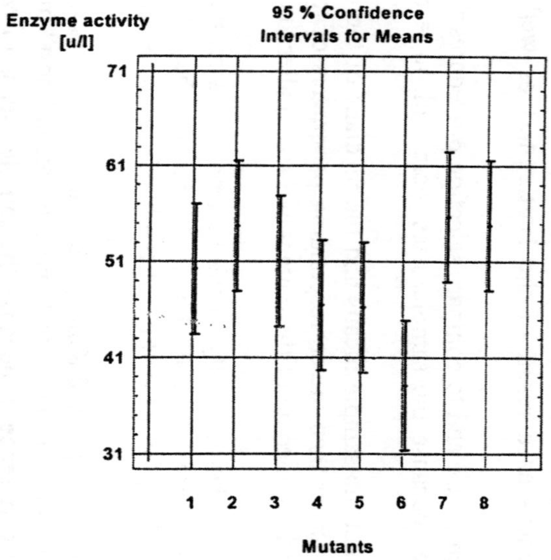

Fig.1. Confidence intervals ($\alpha = 0.05$) for the means of the enzyme activities of the mutants

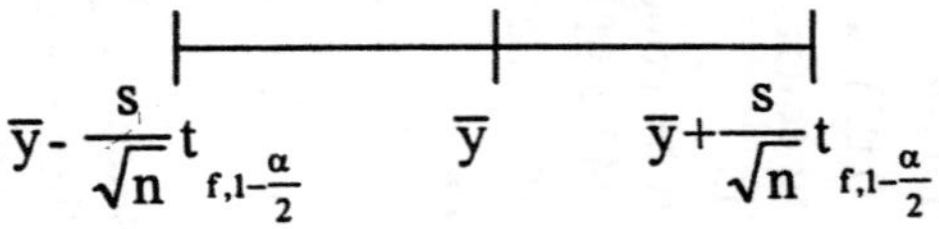

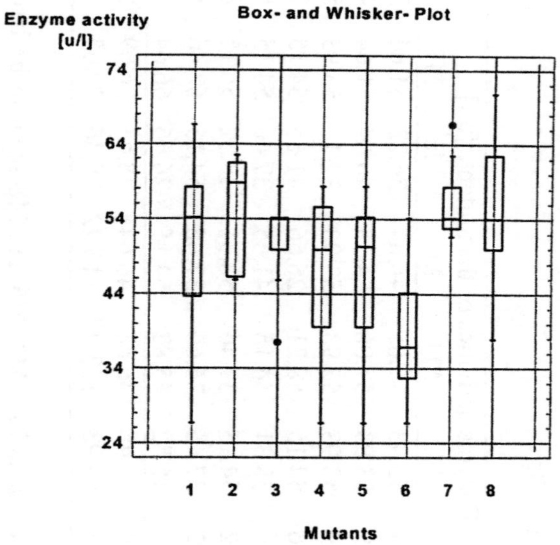

Fig.2. Box- and Whisker-Plots for the enzyme activities of the mutants

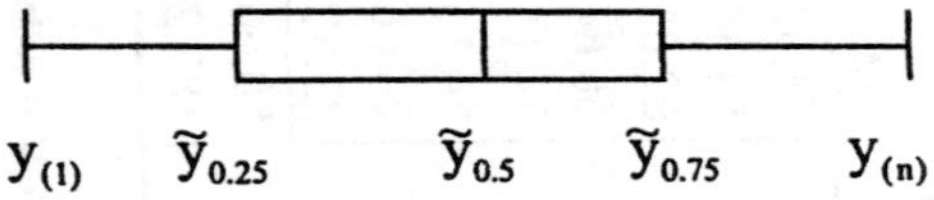

1. Application of the subset selection procedure of Gupta:

Parameters: $k = 8$, $n = 8$, $1\text{-}\beta = 0.90$, $f = 8(8\text{-}1) = 56$, $\rho = 0.5$, $s^2 = 91.93$

$\rightarrow \quad t = t(k\text{-}1; f; \rho; 1\text{-}\beta) = t(7; 56; 0.5; 0.90) = 2.06.$

Procedure: $\bar{y}_i \geq \max_j \bar{y}_j - ts\sqrt{2/n} = 56.09 - 9.88 = 46.21 \qquad i=1,\ldots,8$

Result: On the basis of sample sizes of n = 8 the subset of mutants containing the mutant with the highest enzyme activity with a probability of (1-β) = 0.9 consists of seven elements; all mutants, but M6. For further investigations the experimenter decided to renounce on the mutant M6. That's why the following d-correct procedure refers to k = 7 populations only.

If the samples would base on sizes of n = 20 for instance the subset would include only the mutants with sample means not less than 49.02, that means the five mutants M1, M2, M3, M7 and M8 from the total number of 8.

2. Application of the selection procedure of Bechhofer, Dunnett and Sobel:

Parameters: $k = 7$, $1\text{-}\beta = 0.90$, $f = 7(8\text{-}1) = 49$, $\rho = 0.5$, $s^2 = 94.16$, $d = 4$

$\rightarrow \quad t = t(k\text{-}1; f; \rho; 1\text{-}\beta) = t(6; 49; 0.5; 0.90) = 2.02.$

Procedure: $n = \max\left\{ n_0 ; \text{int}\left(2\left(\frac{s}{d}t\right)^2 + 1\right) \right\} = \max\{8; 49\} = 49.$

Result: For obtaining a precision of d = 4 a sample size of 49 would be necessary to select a good mutant with a probability of (1-β) = 0.9. This means that further 41 sample values must be taken from the seven populations to come to a decision. On the basis of this 49 values the sample means are calculated and the mutant with the largest sample mean will be selected as a „good" mutant with a probability of 0.90.

If a relevant difference of d = 10 would be sufficient for selecting a good population no further experiments would be necessary. On the basis of the eight sample values the mutant M7 with the largest sample mean $\bar{y}_7$ = 56.09 [u/l] would be selected as a good one. Tab. 2 shows the dependence of the necessary sample size on the given precision d.

Tab. 2. Sample sizes for the selection procedure of Bechhofer, Dunnett and Sobel for different precessions d

d	2	3	**4**	5	6	7	8	9	**10**
n	193	86	**49**	31	22	16	13	10	**8**

3. Application of the selection procedure of Gupta and Kim:

Parameters: $k = 7$, $1\text{-}\beta = 0.9025 \rightarrow \sqrt{1-\beta} = 0.95$, $f = 7(8\text{-}1) = 49$, $\rho = 0.5$, $d = 4$,

$\rightarrow t = t(k\text{-}1;\, f;\, \rho;\, \sqrt{1-\beta}\,) = t(6;\, 49;\, 0.5;\, 0.95) = 2.355$, $s^2 = 94.16$.

Procedure: $\bar{y}_i(8) \geq \max_{1 \leq j \leq k} \bar{y}_j(8) - \max\{\,0\,;\, t\,s\,\sqrt{2/n_0} - d\,\} = 48.66$

$i=1,\ldots,k$.

$\rightarrow$ The mutants M1, M2, M3, M7 and M8 could be selected in a first step.

$n = \max\{\, n_0\,;\, \text{int}\,(2\left(\frac{s}{d}t\right)^2 + 1)\,\} = \max\{8\,;\, 66\} = 66.$

Result: In a first step the mutants M1, M2, M3, M7 and M8 could be selected. For coming to a decision between them, further 41 experiments with the mutants would be necessary. On the basis of a sample size of n = 66 the „best“ mutant could be selected then with a probability of (1-β) = 0.9. Tab. 3 contains the necessary sample sizes for different indifference zones d.

Tab. 3. Results of the application of the selection procedure of Gupta and Kim for different indifference zones d

d	**n_0**	**step 1 selection of mutants**	**n**	**step 2 selection of the „best“ mutant**
3	8	M1,M2,M3,M7,M8	117	
4	**8**	**M1,M2,M3,M7,M8**	**66**	
5	8	M1,M2,M3,M7.M8	42	
6	8	M2,M3,M7,M8	30	
8	8	M2,M7,M8	17	
10	**8**	**M7**	**11**	

4. Application of the selection procedure of Dudewicz and Dalal:

Parameters: $k = 8$, $1\text{-}\beta = 0.9$, $h(k,n_0,\beta) = 3.4$, $d = 4$

Procedure: $n_i = \text{int}((h/d)^2 s_i^2 + 1)$

$\rightarrow n_1 = 136$, $n_2 = 43$, $n_3 = 32$, $n_4 = 90$, $n_5 = 84$, $n_6 = 56$, $n_7 = 22$, $n_8 = 73$.

Result: To guarantee a precision of $d = 4$ further $n_i - n_0$ ($i=1,...,5,8$) sample values must be taken from the seven populations for selecting the mutant with the highest enzyme activity with a probability of $(1\text{-}\beta) = 0.9$. On the basis of samples of these sizes n_i the sample means are calculated. Then the mutant with the largest sample mean will be selected as a „good" one with a probability of 0.9. Tab. 4 contains the necessary sample sizes depending on the differences d and the empirical variances. Mutants with large variances in their enzyme activities are „punished" by a large sample size.

Tab. 4. Sample sizes for the selection procedure of Dudewicz and Dalal

d	n_1	n_2	n_3	n_4	n_5	n_6	n_7	n_8
3	241	75	57	159	150	99	39	130
4	136	43	32	90	84	56	22	73
5	87	27	21	58	54	36	14	47
6	61	19	15	40	38	25	10	33
7	45	14	11	30	28	19	7	24
8	34	11	8	23	21	14	6	19
9	27	9	7	18	17	11	5	15
10	22	7	6	15	14	9	4	12

4. Results and Discussion

The application of some statistical selection procedures was illustrated by the example of existing samples of sizes $n = 8$ of the enzyme activities of eight mutants (Tab.1).

Under the supposition of homogeneity of the variances the application of the subset procedure of Gupta provided a subset with seven mutants where the probability that one of the mutants has the highest enzyme activity is 90%. To obtain a smaller subset larger sample sizes would be necessary. To select one „good" mutant the d-correct procedure of Bechhofer, Dunnett and Sobel was used. All mutants whose expected enzyme activities deviated from the largest by less than the given relevant difference $d = 4$ [u/l] were defined as „good" populations. First the procedure calculated the necessary sample size of $n = 49$. Then the sample means had to be calculated, and the mutant with the largest mean will be selected as a „good" one with a probability of 90%.

The procedure of Gupta and Kim calculated a necessary sample size of 66, but in the first step of the procedure the number of mutants had been reduced to five. This procedure should not be used because there were some nearly equivalent good mutants. As long as the variances are assumed to be different, the selection procedure of Dudewicz and Dalal must be used. The sample sizes calculated by the procedure are ranked by the largeness of the variances of the mutants. To guarantee a precision of d = 4 [u/l], further experiments (Tab. 3) would be necessary to select a „good“ mutant with a probability of 90%.

References

Bechhofer, R.E. (1954). A single-sample multiple decision procedure for ranking means of normal populations with known variances. Ann. Math. Stat. **25**, 16-39.

Bechhofer,R.E., Dunnett, C.W.and Sobel, M. (1954). A two-sample multiple decision procedure for ranking means of normal populations with a common unknown variance. Biometrika **41**, 170- 176.

Chiu, W.K. (1974). The ranking of means of normal populations for a generalised selection goal. Biometrika **61**, 579- 584.

Dudewicz, E.J. and Dalal, S.R. (1975). Allocation of observations in ranking and selection with unequal variances. Sankhya Ser.B **37**, 28-78.

Gargova, S. and Tschiriska, J. (1984). Study of the effects of different mutagenic factors on Aspergillus niger. Sci. Rep of the HIFFI Vol XXXI/1, 255-262.

Gupta, S.S. and Kim, W. (1984). A two-stage elimination type procedure for selecting the largest of several normal means with a common unknown variance, in: Design of Experiments: Ranking and selection. Ed. by T.J.Santner and A.C.Tamhane, M.Dekker Inc., New York.

Gupta, S.S. and Panchapakesan, S. (1985). Subset selection procedures: review and assessment. Amer. J. Math. Managem. Sci. **5**), 235-223.

Horn, M. and Vollandt, R. (1995). Multiple Tests und Auswahlverfahren. Reihe Biometrie. G. Fischer Verl. Stuttgart.

Hsu, J.C. (1980). Robust and nonparametric subset selection procedures. Commun. Statist.Theor. Meth. **9**, 1439-1459.

Hsu, J.C. (1981). Simultaneous confidence intervals for all distances from the "best". Ann. Statist. **9**, 1026- 1034.

Rasch, D., Herrendörfer, G., Bock, J., Victor, N. and Guiard, V. (1996). Verfahrensbibliothek- Versuchsplanung und- Auswertung, Bd.1, R. Oldenbourg Verl., München.

Sobel,M. (1967). Nonparametric procedures for selecting the t populations with the largest α- quantile. Ann. Math. Statist. **38**, 1804- 1816.

Modeling and Computation in Pharmaceutical Statistics when Analysing Drug Safety

LUTZ EDLER

Biostatistics Unit - 2070, German Cancer Research Center,
Im Neuenheimer Feld 280, D-69120 Heidelberg, Germany
e-mail: edler@dkfz-heidelberg.de

Key Words:
pharmaceutical statistics, toxicity, safety, stochasticity, longitudinal

Abstract:
Traditional descriptive analysis of drug safety is contrasted with a new approach of modeling the occurrence of adverse events as a time-dependent and stochastic process for each individual patient when clinical studies are performed in pharmaceutical research projects. Safety observed as categorical outcome is modelled as state space process and occurrence as transition between states which enables the application of time-to-event models in the case of a one-dimensional time endpoint or the application of longitudinal models in the case of repeated measurements. The approach is illustrated by the results from a randomized clinical trial on the effects of a bis-phosphonate (Clodronate) in patients with multiple myeloma.

1. Introduction

Drug safety analysis is a part of modern drug development which is almost as important as drug efficacy evaluation. In early clinical trials (Phase I and II) the major goal of safety evaluations is screening, understanding of mechanisms of action and guidance to modifications of applications including decisions on stopping ongoing projects. In later phases (Phase III and especially Phase IV) when an increasing amount of data from patients became available emphasis is on quality control, optimization of drug application and definition of the most

*Kitsos, C.P. and Edler, L., Eds.:Industrial Statistics, Physica-Verlag, Heidelberg, 1997

appropriate target population. Efficient drug safety monitoring systems have been implemented by drug companies for reporting "un-safety" events the socalled adverse events (AEs). In the past, analysis of drug safety was based predominantly on descriptive statistical methods as e.g. frequencies of the occurence of events. This simple analysis is in contrast to the accurate, thorough and costly collection of safety data and also to the elaborate statistical inference techniques used in efficacy analysis.

Realizing that adverse drug effects are observed over time, may last some time and may disappear later it is evident that safety analysis should based on the individual safety process of each individual. This process is intrinsically time-dependent and stochastic: $X_i(t)$, $t > 0$, where t denotes the time since start of treatment and where $i=1, ..., I$ counts the patients. The following sections aim to show consequences of this view for the practice of designing and evaluating clinical safety data. It is demonstrated how traditional analyses fit into this stochastic thinking and how recently proposed methods for longitudinal data provide new tools for the statistical analysis.

2. Safety Described as Stochastic Process

Primarily, safety data are either categorical obtained by a toxicity (see e.g. the CTC criteria, Berdel et al, 1994) or continuous laboratory measurements. The latter measurements are often transformed into the standard categories *0 = normal range, 1 = mildly (elevated/reduced), 2 = moderately, 3 = severe, 4 = life threatening*. Therefore, without substantial restriction of generality it is assumed for the following that safety variables are recorded on an ordinal measurement scale with values in a state space $S = \{0, 1, 2, 3, 4\}$. Descriptive safety statistics consider the frequencies $f(s)$ of cases in these safety classes over an observation period. Relative proportions are calculated as ratios $f(s)/n$ where n is the basic number of evaluable patients. An example is given in Table 1 from the data analysis of a clinical trial.

Table 1: Descriptive presentation of hematological adverse events with respect to the levels of hemoglobine (HB) in patients with multiple myeloma who were studied in a multicenter randomized clinical trial on the efficacy of a bis-phosphonate (Clodronate) for improvement of bone impairment. Both, for the treatment group (CLODRONATE) and the control group (CONTROL) the absolute and the relative frequencies of patients in the five WHO categories are exhibited at the time of start of treatment (BASELINE) at the end of treatment (FINAL) and during treatment by reporting the highest category seen (WORST CASE).

	CLODRONATE WHO - Categories					CONTROL WHO - Categories				
	0	1	2	3	4	0	1	2	3	4
BASELINE	44	13	9	4	1	42	19	10	1	0
%	*62.0*	*18.3*	*12.7*	*5.6*	*1.4*	*58.3*	*26.3*	*14.1*	*1.4*	-
FINAL	32	16	17	3	3	42	18	7	3	2
%	*45.1*	*22.5*	*23.9*	*4.2*	*4.2*	*58.3*	*25.0*	*9.7*	*4.2*	*2.8*
WORST	15	20	17	10	9	18	25	19	6	4
%	*21.1*	*28.2*	*23.9*	*14.1*	*12.7*	*25.0*	*34.7*	*26.3*	*8.3*	*5.5*

This analysis does not account for the four-weekly schedule of adverse event recording, the time of the appearance or disappearance of the adverse events, and the individual toxicity burdon over the time. Therefore, it is more appropriate to consider the occurrence of those events over time as a stochastic adverse event process $\{X(t, w), t \geq 0\}$ for each patient w with values in a finte state space S. Notice, $X(t, w) = s$ says simply that patient w has at time t the adverse event status $s \in S$. $X(t, w)$ is defined for all times t, but observations are only available a for a limited number of discrete time points t_j, say $j = 1, ..., n_i$.

Whereas time-dependency is the result of the prolonged observation of a patient, stochasticity is the result of the inter-individual variability of the responses of the patients. Major sources of this variability are the disease, the modifications of the investigation protocol and the medical examiner who judges safety. An investigation of Brundage et al 1993 found that factors relating to the examiner and to the examination contribute more to observer variability than do the operational criteria of definitions. Even experienced examinors draw variable conclusions with considerable intra-rater variability. Those variations can hardly be modeled and therefore contribute in most analyses directly to the error term. Also the data manager and the documentation systems may contribute to variability. This all together constitutes random variation which is at best modelled statistically by a random variable.

Modeling safety as a stochastic process does not impede traditional descriptive analysis since the commonly used measures can be defined within this frameworkas e.g.

$M_k = \underset{t>o}{Max}\ X_k(t)$ the maximum value under treatment,

$E_k = \underset{t>0}{\inf}\{X_k(t) \geq B\}$ the first entrance time into an adverse region,

$\underline{M}_k = M_k - X_k(0)$ the maximum over individual 'background',

$m_k = \underset{t>0}{Median}\ X_k(t) / M_k$ the median compared with the maximum toxicity

$AUC\text{-}TOX = \int_0^t X_k(u)du$ the cumulative toxicity by time t.

In general, a safety index is a functional F of the safety process X(t):

$F = F\{X(t), t \geq 0\}$.

3. State Specific Safety Analysis

State specific models have been developed within the framework of Markov Processes and applied in disease state models; (Chiang 1975). Figure 1 shows the state space and the transitions between states for the evaluation of one categorial adverse event variable X (t). Three additional states may be relevant in praxis: the two absorbing states „death“ and „drop-out“ and the transient state of „intermittent missing“ values. Their consideration enlarges the previously defined state space $S = \{0, 1, 2, 3, 4\}$ to $S = \{0, 1, 2, 3, 4, 5, 6, 7\}$.

Synchronous Discrete Time Observation Scheme: Markov Chain. Cyclic treatment or cyclic inspection schemes allow often the assumption of a synchronous discrete schedule of times $0 = t_0 < t_1 < t_2 < ... < t_n$ at which each patient is assessed. Hence, let us assume that the adverse event variable X is observed at fixed discrete time points t_n for patient (i) with values $X^{(i)}(t_j) = Y_{ij}$ for $j = 0, 1, ..., n_i$. Such an observation scheme applies when patients are examined according to exactly the same time schedule, e.g. every day, every Monday each week etc. The only variation allowed is a different length n_i of the observation scheme. When the transition from state s_{j-1} at time t_{j-1} into state s_j at time t_j does not depend from previous visits at times t_{j-2}, t_{j-3}, ..., X(t) is Markovian, Given the the measurements $Y_j = X(t_j)$ and a non-informative initial distribution p_0 then a Markov chain is defined by its transition probability matrix π on S x S with elements

(1) $$\pi_{ab} = P(Y_i = b \mid Y_{i-1} = a), \qquad a, b \in \{0, ..., 7\}.$$

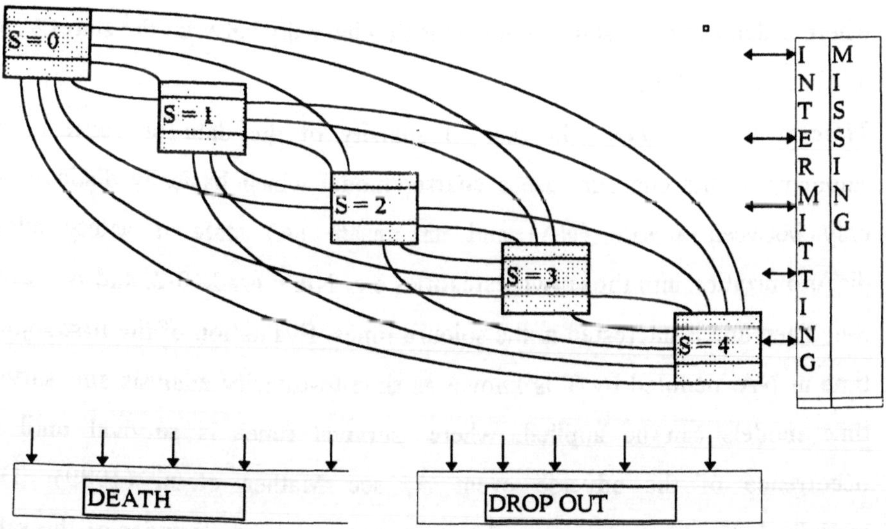

Figure 1: State space model and its transitions when the categorical classification into the 5 WHO levels are used and when death and drop-out are further absorbing states and intermitting missing is a further transient state.

The transition probabilities π_{ab} are estimated by the relative frequencies h_{ab}/h_a where h_{ab} is the number of transitions observed between the states a and b and where h_a is the total number of transitions starting from state a, both counted over all times $t_j \geq 0$ and over all patients. h_{ab}/h_a is the maximum likelihood estimate if death, drop-out and intermittent missing are non-informative.

Continuous Time Observation Scheme: Counting Process. The Markov chain approach can be generalized to an asynchronous observation scheme by using survival analysis methods. For this an indicator process $Y_{ia}(t)$ is defined

$$(2) \qquad Y_{ia}(t) = \begin{cases} 1 & \text{patient i is in state a at time t} \\ 0 & \text{else} \end{cases}$$

and the data are summarized as counting process $N_{iab}(t)$ = Number of transitions from a to b in the time interval [0,t], a, b, ε S = {0, ..., 7}. The transitions are modeled by the intensity process

$$(3) \qquad \alpha_{ab}(t)\, Y_{ia}(t)$$

where α denotes the instantaneous transition intensity between the states a and b, a hazard type function.

Time-to Event models. In case of sparsity of the data or content with summary results one may use a coarser model which basically distinguishes only between a safe (NA) and an unsafe (A) state if safety allows dichotomization into those two categories e.g. NA = grade 0-2, and A = grade 3-4. Then one is interested in the sojourn times. Evaluation of the first sojourn time in NA, denoted by T is known as time-to-toxicity analysis and survival time models can be applied, where 'survival time' is survival until the occurrence of the adverse event A, see Mather et al (1989). Their modelincludes the actual survival time X as possible confounder of the safety process. Estimates of the crude probability of toxicity by time t

$$(4) \qquad P(T \leq t \text{ and } X > t)$$

are obtained from the two net probabilities of toxicity and of survival

(5) $F^{T|S}(t) = P(T \leq t \mid X > t)$ and $S^{S|T}(t) = P(X > t \mid T > t)$.

4. Longitudinal Models for Safety Analysis

Treatment and observation lasts in clinical trials of chronic diseases usually several months and years such that a number of repeated measurements of safety variables are available for evaluation. Recently proposed methods of categorial longitudinal data elaborated e.g by Liang et al (1992) and Diggle et al (1994) became useful for clinical safety analyses. At first, assume that safety monitoring yielded synchronous observations for all patients

$$Y^{(i)} = (Y_{i1},\ldots,Y_{in}),\ Y_{ij}\ \varepsilon\ \ S,\ i=1,\ldots I.$$

Would have been the safety variable continuous - which may happen when a quantitative laboratory parameter is evaluated - then the standard linear model for multivariate data would be applicable at least as long as a multivariate normal distribution is assumed, see e.g. Diggle et al (1994, Chapter 4 and 5. Summation over a number of safety scores, possibly weighted, could provide such an endpoint even if the observations are categorical. The analysis of single categorial safety data requires generalized linear longitudinal models, where distinction has been made between

(i) *marginal models*

(ii) *transition models*

(iii) *random effect models.*

Marginal models are appropriate when the relationship between the covariates and the marginal probabilities of the adverse event categories is of primary interest. The marginal expectation

(6) $$E[Y_{ij}] = \mu_{ij}$$

is modelled via a link

(7) $$h(\mu_{ij}) = z_{ij}\beta$$

The marginal variance may depend on the marginal means. The within-patient correlation is a function of the marginal means and additional model parameter.

In contrast to marginal modeling - *random effects models* will not be considered further- *transition models* have the advantage that they allow past safety values influence present safety. Generalized estimating equations have been aplied successfully by Lipsitz et al (1994). Synchronized data and a balanced design of safety data was assumed:

$$Y_{it} \qquad i = 1, \dots, N, t = 1, \dots, T,$$

where for each patient (i) measurements were obtained in each of the T time points or time intervals, e.g. treatment cycles. This can be achieved at cost of creating intermittent missing values at some times for some patients and/or at cost of a wider observation mesh. The amount of missing values can be reduced by using a relatively wide mesh of time points and by applying locally carry forward or carry backward methods for imputing missing values. Needless to say, that any such procedure should be fixed before the start of the analysis either in a study protocol or in a statistical evaluation plan. A covariate process z_{it} e.g. describing treatment, can be included into the model. The transition model uses past safety information

(8) $$H_{it} = \{Y_{is}, s=1, \dots t-1\}$$

and safety at time t is then modeled as a conditional generalized linear model

(9) $$f(y_{it} \mid H_{it})$$

where the density f is taken from an exponential family. Using GLIM methodology the expected safety of patient (i) at time t given the past is

(10) $$\mu_{it} = E\,[Y_{it} / H_{it}]$$

which is then linked to the covariate process by an appropriate the link function.

Markov Transition Model. A more direct analysis of categorial safety data with the state space $S = \{0, \dots, 4\}$ from above - not considering the absorbing states and missing information- is obtained with the one-step transition probabilities of a Markov chain

(11) $\pi_{ab} = P(Y_i = b \mid Y_{i-1} = a)$, $a, b \, \varepsilon \, \{0, ..., 4\}$

and the proportional odds model using logistic transformations

(12) $\text{logit } P(Y_{it} \leq b \,/\, Y_{it-1} = a) = \theta_{ab} + z_{it} \cdot \beta_a$

see Diggle et al (1994), Chapter 10. This set of equations can be put into one closed-form equation when dummy variables defined as follows

(13) $$Y^*_{i,t,a} = \begin{cases} 1 & \text{if } Y_{it} = a \\ 0 & \text{else} \end{cases}.$$

Denoting

(14) $$\mathbf{Y}^*_{i,t-1} = (\mathbf{Y}^*_{i,t-1,0},\ \mathbf{Y}^*_{i,t-1,1},\ \mathbf{Y}^*_{i,t-1,2},\ \mathbf{Y}^*_{i,t-1,3})$$

the generalized logistic model is

(15) $$\text{logit}\{P(Y_{it} \leq b | Y^*_{i,t-1} = y^*_{i,t-1})\} = \theta_b + \sum_{m=1}^{3} \alpha_{m,b} y^*_{i,t-1,m} + z_{it}[\beta + \sum_{m=1}^{3} \gamma_m y^*_{i,t-1,m}]$$

for b = 0, 1, 2, 3 . The relation to eq (12) is given by the restrictions $\theta_b = \theta_{4b}$, $\alpha_{ab} = \theta_{ab} - \theta_{a+1b}$ and $\gamma_a = \beta_a - \beta_{a+1}$. Extensions to the Non-Markovian case are possible.

5. Example: A Multicenter Randomized Trial

In this section, the result of the marginal longitudinal model applied to the categorial hematological toxicity variable HB (hemoglobin) will be shown. The objective of the clinical trial was the the evaluation of efficacy and safety of Clodronate treatment of patients with multiple myeloma for the improvement of bone impairment (osteolytic lesions, osteoporosis, fractures), see Heim et al (1995). The descriptive summary on HB has already been given in Table 1. Table 2 shows the results of the marginal model applied according to Lipsitz et al (1994) for two covariance structures of the repeated measurements: independence and exchangability. Use of robust variance estimates for the parameters and a banded correlation structure gave similar results. No significant difference between the two treatment arms was seen ($p > 0.1$), irrespective how the correlation structure between the time points was modelled. Such that it was concluded that the safety with respect to HB was not of major issue of concern in this study.

Table 2: Hemoglobin (HB) toxicity in the two arms of a randomized clinical trial comparing Clodronate with Placebo in myeloma cancer patients by using *the marginal model* according to Lipsitz et al (1994). Results are provided for the model with independent correlation structure and exchangable correlation structure of the repeated measurements. A cumulative logistic model was applied resulting in for intercepts s=0, s≤1, s≤2, s≤3

	INDEPENDENT			EXCHANGEABLE		
	$\hat{\beta}$	s.e.	p-value	$\hat{\beta}$	s.e.	p-value
Int. $S = 0$	-0.17	(0.18)	0.4	-0.37	(0.37)	0.3
Int. $S \leq 1$	0.97	(0.19)	< 0.1	0.75	(0.36)	0.04
Int. $S \leq 2$	2.34	(0.21)	< 0.1	2.10	(0.38)	< 0.1
Int. $S \leq 3$	3.64	(0.28)	< 0.1	3.44	(0.41)	< 0.1
Time(months)	0.079	(0.02)	< 0.1	0.015	(0.011)	0.18
Treatment	0.17	(0.11)	0.12	0.32	(0.23)	0.16

The analysis was performed with the SAS Macro MULTRPM for analyzing clustered categorial data based on the GEE approach of Lipsitz et al (1994).

6. Discussion

Safety data are collected and processed in clinical trials in order to assess the hazard of drug treatment and to achive best possible protection of the patient. Common to safety data is the time dimension. This provides the basis to consider safety as a stochastic process X(t)in time. The subject of this paper was to elaborate the consequences of this concept for an improved safety evaluation in clinical trials. It was shown that stochastic modeling of the safety process provides new insights and access to efficient statistical analysis tools. Naturally, restrictions had to made both by assumptions about the data and the methods applied. A detailed analysis was presented for synchronized and balanced data und the example was restricted to the marginal model. Other models, as the Markov transition model, were outlined. In all these cases it is essential to model the dependencies between the repeated observations.

Not considered here are methods for the assessment of *rare adverse events* in observational studies (Feldmann, 1993). One should also remark that a safety evaluation is highly *multidimensional:* WHO considers about p = 20 items (WHO, 1979) and the CTC (Common Toxicity Criteria) of the NIH consider at present more than 100 adverse event types (Green&Weiss, 1992). Safety evaluation is always affected by *length of survival* as mentioned in the time-to-event models. Intercurrent mortality -not related to toxicity- may prevent the observation of toxicity which is a well known confounding effect. Problems with drop out and missing values were not further discussed above, but see Diggle et al (1994) and Diggle & Kenward (1994). The *pre-treatment time interval* which defines the baseline values for a safety analysis has to be investigated better and safety analysis should account for the course of the process earlier than the start of treatment (t=0). The safety process is affected by treatment schedule and concommittant treatment. When the treatment is administered in *regular cycles* the safety process may respond to this schedule as a cyclic process $X(r\tau + t) = X(t)$ for $0 \leq t \leq \tau$ when r counts the cycles and τ is the length of the cycle. *Concommittant treatment* may may increase toxicity e.g. by a synergistic mode of action, or it may decrease toxicity if the treatment is given for prevention of toxicity. Assessment of *causal* relationships between treatment and safety would augment the state space S, e.g to a product state space.

References

Berdel WE, Becher R, Edler L, Hanauske AR., Kerpel-Fronius S, Rauschning W, Schmoll HJ, Sindermann H (1994): Standardarbeitsanweisungen (Standard Operating Procedures), SOP) derPhase I/II Studiengruppe der Arbeitsgemeinschaft für internistische Onkologie (AIO) in der Deutschen Krebsgesellschaft. Onkologie, 17, 311-338.

Brundage MD, Pater JL, Zee B (1993): Assessing the reliability ot two toxicity scales: Implication for interpreting toxicity data. J. Natl. Cancer Inst. 85, 1138.

Chiang CL (1975): An Introduction to Stochastic Processes and Their Applications. R.E. Krieger : Huntington, New York.

Cook RJ., Farewell VT. (1994): Guidelines for monitoring efficacy and toxicity responses in clinical trials. Biometrics 50, 1146-1152.

Diggle PJ, Liang K-Y, Zeger SL (1994): Analysis of Longitudinal Data. Clarendon Press: Oxford.

Diggle P, Kenward MG (1994): Informative drop-out in longitudinal data analysis (with discussion). Appl. Statist. 43, 49-93.

Feldmann U (1993): Design and analysis of drug safety studies, with special reference to sporadic drug use and acute adverse reactions. J. Clin Epidemiol 46, 237-243.

Green, S., Weiss, G.R. (1992): Southwest Oncology Group standard response criteria, endpoint definitions and toxicity criteria. Investigational New Drugs 10, 239-253.

Heim ME, Clemens MR, Queisser W, Pecherstorfer M, Boewer H, Herold F, Franke A, Hermann Z, Loose R, Edler L (1995): Prospective randomized trial of dichloromethylene bisphosphonate (Clodronate) in multiple myeloma requiring treatment. A Multicenter Study. Onkologie 18, 439-448.

Lipsitz SR, Kim K, Zhao L (1994): Analysis of repeated categorial data using generalized estimating equations. Stat. Med. 13, 1149-1163.

Mather FJ, Simon RM, Clark GM, Von Hoff DD)1989): A method for the evaluation of dose-toxicity relationships in clinical trials. Statistics in Medicine 8, 813-823.

Miller AB, Hoogstraten B, Staquet M, Winkler A (1981): Reporting results of cancer treatment. Cancer 47, 207-214.

WHO (1979): Handbook for Reporting Results of Cancer Treatment. WHO Offset Publication No 48, World Health Organization: Geneva.

Isotonic Inference with Particular Interest in Application to Clinical Trials

C. HIROTSU
Faculty of Engineering, University of Tokyo
Hongo, Bunkyoku, Tokyo, Japan

Key Words: Ordered alternative, Ordered categorical data, Cumulative efficient scores

Abstract

Various isotonic hypotheses are introduced in comparative clinical trials including a monotone dose-response relationship and its extension to two-way problems. Then a systematic approch for testing those hypothese based on the cumulative efficient scores is given with some applications to real data.

1 Introduction

Isotonic inference has many applications in comparative clinical trials.

First there are many interesting isotonic hypotheses associated with the dose-response relationship. One general approach to the problem is to assume a parametric model and make an inference on the parameters describing it. The inference based on the parametric model can be, however, wildly biased and variable when the specified model is incorrect. It has been pointed out in the environmental toxical applications, for example, that no parametric dose-response model can be assumed to hold generally at very low doses of interest and yet monotone relationship might be assumed and give a more reliable result. Further it also sometimes suffices to show an overall upward trend or to detect a steep change-point in the responses. It is then unnecessary to assume a rigid parametric model, and a nonparametric trend test or some multiple comparisons procedure assuming only a monotone relationship is more appropriate.

*Kitsos, C.P. and Edler, L., Eds.:Industrial Statistics, Physica-Verlag, Heidelberg, 1997

The second problem considered is associated with the analysis of a comparative clinical trial based on the ordered categorical responses. Suppose that there are two treatments evaluated in k ordered categories. Denoting the occurrence probability of (i, j) cell by p_{ij} we are interested in testing the null hypothesis $H_0 : p_{ij} = p_{i.}p_{.j}$ that there is no treatment effect. The usual goodness of fit chi-squared is, however, inappropriate since we are interested in a particular (retricted) alternative that either of the two treatments is superior to the other.

We develop a general approach to testing such ordered alternatives based on the cumulative efficient scores. A relationship of the approach to the change-point analysis is also mentioned.

2 The case for isotonic inference

The data in Table 1 are measurements of the half-life of an antibiotic drug in relation to the dose administered. The usual analysis of variance (ANOVA) is obviously inappropriate, because of the ordering of the doses, and one possible approach is to assume a parametric model. The simplest model for the monotone relationship is a linear regression model. It is, however, generally difficult to assume that a linear relationship holds over a wide range of an explanatory variable. For the dose-response relation there are of course more natural response curves such as a sigmoid function, but it is still often very hard to assume a particular model for the given set of data. Our approach then assumes only a monotone relationship in the mean half-life,

$$H_1 : \mu_1 \leq \cdots \leq \mu_a \,, \tag{1}$$

where at least one inequality is strong, so that the null model $H_0 : \mu_1 = \cdots = \mu_a$ is excluded.

Table 1 Half life of an antibiotic in rats

Dose (mg/kg)	Data (hr)					Average
5	1.17	1.12	1.07	0.98	1.04	1.076
10	1.00	1.21	1.24	1.14	1.34	1.186
25	1.55	1.63	1.49	1.53		1.550
50	1.21	1.63	1.37	1.50	1.81	1.504
200	1.78	1.93	1.80	2.07	1.70	1.856

The data in Table 2 show ordinal categorical data typical of a phase III comparative clinical trial. Assuming a multinomial model with cell probabilities p_{ij}, the null hypothesis that the two treatments are equal can be expressed as $p_{1j} = p_{2j},\ j = 1, \cdots, 4$, or equivalently as

$$p_{ij} = p_{i.}\, p_{.j}. \tag{2}$$

The equation (2) is the familiar independence hypothesis for a two-way contingency table. However, the usual goodness of fit chi-square test is inappropriate since we are interested in a more restricted alternative

$$H_2: \begin{array}{ll} p_{11}/p_{21} & \le \cdots \le p_{14}/p_{24} \\ \text{or} & \\ p_{11}/p_{21} & \ge \cdots \ge p_{14}/p_{24}, \end{array}$$

where at least one inequality is strong, implying that treatment 1 is superior to treatment 2 in efficacy or vice versa. Ordinal categorical data are a special case of rank data with many ties, and any method for rank data can be applied to ordinal categorical data, and vice versa.

Table 2 Efficacy in a phase III trial of antibiotics

Drug	Not effective	Slightly effective	Effective	Excellent
AMPC	3	8	30	22
S6472	8	9	29	11

If ordinal categorical data are obtained at several doses, as in Table 3, we are interested in testing the two-way ordered alternative

$$H_3: p_{i+1j}/p_{ij} < p_{i+1j+1}/p_{ij+1}, \quad i = 1, \cdots, a-1; j = 1, \cdots, b-1,$$

which implies that higher doses are superior to lower doses in efficacy.

Table 3 Usefulness in a dose finding clinical trial

Drug	Undesirable	Slightly undesirable	Not useful	Slightly useful	Useful	Excellent
Placebo	3	6	37	9	15	1
AF 3 mg/kg	7	5	33	21	10	1
AF 6 mg/kg	5	6	21	16	23	6

3 Various extensions of the monotone relationship

A monotone dose-response relationship may be disturbed by toxity at higher doses, and a nonparametric testing procedure for the downturn hypothesis

$$H_4: \mu_1 \le \cdots \le \mu_{\tau+1} \ge \mu_{\tau+2} \ge \cdots \ge \mu_a, \quad \tau = 1, \cdots, a-1,$$

has been proposed in Simpson and Margolin (1986), where τ is an unknown turning point.

Some other extensions arise when responses show monotone relationships with the passage of time. Frequently encountered examples include the monotonic change of occurrence probabilities of some events, increasing treatment effects, and increasing hazard rates with time. For instance, the hypothesis

$$H_5 : \mu_2 - \mu_1 \le \mu_3 - \mu_2 \le \cdots \le \mu_a - \mu_{a-1} \tag{3}$$

arises from the analysis of an age-period-cohort effects model where only the second order differences are estimable in each effect along with the time axis, see Hirotsu (1988). The hypothesis H_5 is equivalent to $H_5' : \mu_i - 2\mu_{i+1} + \mu_{i+2} \ge 0$, and may be called a 'convexity hypothesis'.

As seen from the above examples, the isotonic inference is closely related to the change-point analysis. Actually a one-sided change-point model may be formulated as a set of particular monotone relationships,

$$H_6 : \mu_1 = \cdots = \mu_\tau < \mu_{\tau+1} = \cdots = \mu_a, \quad \tau = 1, \cdots, a-1, \tag{4}$$

with τ an unknown change-point parameter, so that a useful statistic for change-point analysis is useful also for isotonic inference. Interestingly, (4) defines $a-1$ edges of the convex cone defined by the simple ordered alternative (1).

4 A complete class lemma

Suppose that $\mathbf{y}$ is distributed as an a-variate normal with mean vector μ and known covariance matrix $\mathbf{\Omega}$. Noting that the hypotheses (1) and (3) can be expressed in the matrix notation as $K_1 : A'\mu \ge \mathbf{0}$, we consider generally testing the null hypothesis $K_0 : A'\mu = \mathbf{0}$ against the alternative K_1, where the inequality is element-wise with at least one inequality strong. Then the essentially complete class of the test is formed from all those convex acceptance regions that are increasing in every element of

$$(A'A)^{-1}A'\Omega^{-1}[\mathbf{y} - \mathrm{E}_0(\mathbf{y})], \tag{5}$$

where $\mathrm{E}_0(\mathbf{y})$ is the conditional expectation of $\mathbf{y}$ given the sufficient statistics under the null hypothesis H_0. In a more general situation that $\mathbf{y}$ is distributed according to some distribution with the unknown parameter vector μ with possibly some nuisance parameters, the asymptotically equivalent of the equation (5) is

$$(A'A)^{-1}A'\nu(\hat{\mu}_0), \tag{6}$$

where $\nu(\hat{\mu}_0)$ is the efficient score vector evaluated at the maximum likelihood estimator $\hat{\mu}_0$ under the null hypothesis, see Hirotsu (1982) and Cohen & Sackrowitz (1991) for details. The equations (5) and (6) can sometimes be very simple and give reasonable test statistics.

5 Testing a simple ordered alternative in the normal means

5.1 Test statistics

As an important special case we consider the simple ordered alternative in the one-way layout,

$$y_{ij} = \mu_i + \varepsilon_{ij},\ i = 1, \cdots, a;\ j = 1, \cdots, n_i,$$

where the ε_{ij} are assumed to be independently distributed as normal $N(0, \sigma^2)$ with the known variance σ^2. Then the formula (5) can be applied with $\mathbf{y} = (\bar{y}_{1.}, \cdots, \bar{y}_{a.})', \mathbf{\Omega} = \mathrm{diag}(\sigma^2/n_i)$ and $A' = D_a'$ an $a-1 \times a$ differential matrix with the ith row $(\,0 \cdots 0 - 110 \cdots 0)$. The ith standardized element of $(D_a' D_a)^{-1} D_a' \Omega^{-1}[\mathbf{y} - \mathrm{E}_0(\mathbf{y})]$ is

$$y_i^* = \frac{1}{\sigma}\left(\frac{1}{N_i} + \frac{1}{N_i^*}\right)^{-1/2}(\bar{Y}_i^* - \bar{Y}_i),\ i = 1, \cdots, a-1, \tag{7}$$

where $N_i = n_1 + \cdots + n_i,\ N_i^* = n_{i+1} + \cdots + n_a$ and

$$\bar{Y}_i = (y_{1.} + \cdots + y_{i.})/N_i,\ \bar{Y}_i^* = (y_{i+1.} + \cdots + y_{a.})/N_i^*. \tag{8}$$

According to the complete class lemma there have been proposed two statistics,

the cumulative chi-squared	:	χ^{*2}	=	Σy_i^{*2},
the maximal component of χ^{*2}	:	$\max t$	=	$\max y_i^*$.

The χ^{*2} is for the two-sided alternative and $\max t$ for the original one-sided hypothesis H_1. The $\max t$ is easily extended to the two-sided problems based on the maximum of y_i^{*2}.

The very characteristics of the χ^{*2} and max t are the directional properties deduced from the strong positive correlations among y_i^*'s. In particular the χ^{*2} in the balanced case is expanded in the series of independent chi-squared variables,

$$\chi^{*2} = \frac{a}{1 \cdot 2}\chi^2_{(1)} + \frac{a}{2 \cdot 3}\chi^2_{(2)} + \cdots + \frac{a}{(a-1) \cdot a}\chi^2_{(a-1)},$$

where $\chi^2_{(l)}$ is the one df chi-squared detecting the departure from the null model in the direction of Chebyshev's lth order orthogonal polynomial. Hence the χ^{*2} is testing mainly but not exclusively a linear trend, see Hirotsu (1986) and Nair (1986) for details. The statistic $\max y_i^{*2}$ is also characterized as the likelihood ratio test statistic for the change of means in the independent sequence of normal variables. For the similar arguments for the convexity hypothesis H_5 one should refer to Hirotsu (1986, 88).

5.2 Distribution theory

There is a very good chi-squared approximation for the null and alternative distributions of the χ^{*2}. Under the null model the χ^{*2} is well approximated by a constant times chi-squared $d\chi^2(f)$ with constants

$$\left\{\begin{array}{lcl} d & = & 1+\frac{2}{a-1}(\frac{\lambda_1}{\lambda_2}+\frac{\lambda_1+\lambda_2}{\lambda_3}+\cdots+\frac{\lambda_1+\cdots+\lambda_{a-2}}{\lambda_{a-1}}), \\ f & = & (a-1)/d, \end{array}\right. \quad (9)$$

where $\lambda_i = N_i/N_i^*$. An even better approximation based on the expansions by Laguérre's orthogonal polynomials and also the approximation under the alternative hypothesis are given in Hirotsu (1979b). For calculating the p value of $\max t$ an exact and very efficient algorithm has been obtained based on the Markov property of the subsequent components $y_i^*, i = 1, \cdots, a-1$ (Hawkins, 1977).

The power of the χ^{*2} and $\max t$ in comparisons with the likelihood ratio test has been discussed on several occasions, see Hirotsu (1978, 79a, 92) and Hirotsu et al. (1992), for example. The χ^{*2} is useful for an overall trend test whereas $\max t$ gives a multiple comparisons procedure useful for detecting a gap in the subsequent means. As such multiple comparisons procedures the Williams' (1971) method and its modification by Marcus (1976) have been proposed for defining the noneffective dose level. Just as the $\max t$ is the maximal component of the χ^{*2} the modified Williams' statistic is the maximal component of the Bartholomew's $\bar{\chi}^2$ test.

5.3 Application of χ^{*2} and $\max t$ to Table 1

Since the variance σ^2 is unknown in (7) it is replaced by the usual unbiased variance $\hat{\sigma}^2 = \Sigma\Sigma(y_{ij} - \bar{y}_{i.})^2/(24-5) = 0.020741$ to define the statistic

$$F^* = \frac{1}{a-1}\chi^{*2}\mid_{\sigma^2=\hat{\sigma}^2} = 54.739.$$

The statistic F^* is well approximated under the null hypothesis by the F distribution with the df $(f, \Sigma n_i - a)$ where $f = 2.067$ by the formula (9). Therefore the p value is evaluated as 1.1×10^{-8}, which should be compared with the p value 5.0×10^{-7} of the usual F test.

The maximal component is obtained at the partition between the levels 2 and 3 like

$$\max t = [(\frac{1}{10}+\frac{1}{14})\cdot 0.02741]^{-1/2}(\frac{23.00}{14}-\frac{11.31}{10}) = 8.584,$$

whose one-sided p value is evaluated exactly as 1.1×10^{-7} .

6 Testing ordered alternatives in Binomial probabilities

Suppose that the y_i are independently distributed as binomial $B(n_i, p_i)$ and we are interested in testing the simple ordered alternative

$$H_1 : p_1 \le \cdots \le p_a.$$

If there are attached to the y_i the quantitative measures $d_1 < \cdots < d_a$ then the locally most powerful test against the wide range of monotone relationship of p_i to d_i is obtained by Cochran (1954) and Armitage (1955). Without the information of d_i the general complete class lemma is applied to obtain

$$\begin{array}{llcl} \text{the cumulative chi-squared} & : \quad \chi^{*2} & = & \Sigma y_i^{*2}, \\ \text{the maximal component of } \chi^{*2} & : \quad \max t & = & \max y_i^*, \end{array} \qquad (10)$$

where y_i^* is given by (7) with σ replaced by $[\bar{Y}(1-\bar{Y})]^{1/2}$, $\bar{Y} = \Sigma y_i / \Sigma n_i$, and $y_{i.}$ replaced by y_i in (8). The formula (9) is also valid for the χ^{*2} and for $\max t$ again an exact algorithm is available based on the Markov property of the y_i^*.

7 Analyzing a two-way contingency table with ordered column categories

7.1 Two sample problem

First consider the two sample problem presented by Table 2. Then a most popular approach will be a nonparametric test based on the linear score statistic such as Wilcoxon's. Now for the two-sided alternative H_2 the two statistics

$$\text{the cumulative chi-squared} \quad : \quad \chi^2 = \chi_1^2 + \cdots + \chi_{b-1}^{*2} \qquad (11)$$

$$\text{the maximal component of } \chi^{*2} \quad : \quad \max \chi^2 = \max \chi_j^2 \qquad (12)$$

is defined according to the complete class lemma based on the goodness of fit chi-squared χ_j^2 for the 2×2 table formed by accumulating the first j and the rest $b-j$ columns, respectively. The χ_j^2 is, however, nothing but the y_i^{*2} of (10) if the binomial data were arranged in $2 \times b$ table in an obvious way and exactly the same distribution theory applies also to this case. The two-sided p values are 0.039 for χ^{*2} and 0.154 for $\max \chi^2$ whereas it is 0.025 for the Wilcoxon test. In this case the Wilcoxon test shows the smallest p value since approximately a linear trend is observed in p_{1j}/p_{2j}, $j = 1, \cdots, 4$. If those tests are applied to the last two rows of Table 3 for comparing AF3mg and AF6mg then the two-sided p values are 0.0128, 0.0096 and 0.0033 for the Wilcoxon, the χ^{*2} and $\max \chi^2$ methods, respectively.

7.2 General a sample problem

For a general a sample problem the Wilcoxon test is extended to the Kruskal-Wallis test. The same type extensions are available for the χ^{*2} and its maximal component by defining the χ_j^2 in (11) and (12) as the goodness of fit chi-squared for the accumulated $a \times 2$ table for the partition between the columns j and $j+1$. The constants for the chi-squared approximation of χ^{*2} are obtained by

$$\begin{cases} d &= 1 + \frac{2}{b-1}\left(\frac{\gamma_1}{\gamma_2} + \frac{\gamma_1+\gamma_2}{\gamma_3} + \cdots + \frac{\gamma_1+\cdots+\gamma_{b-2}}{\gamma_{b-1}}\right) \\ f &= (a-1)(b-1)/d \end{cases}$$

with $\gamma_j = C_j/C_j^*$, $C_j = y_{\cdot 1} + \cdots + y_{\cdot j}$, $C_j^* = y_{\cdot j+1} + \cdots + y_{\cdot b}$. The max χ^2 can be evaluated by the algorithm based on the Markov property of the subsequent χ_j^2's (Hitotsu et al., 1992).

There are also many interesting multiple comparisons procedures and their applications, see Hirotsu (1992) for details.

8 Two-way contingency table with natural orderings in both of rows and columns

Assuming a multinomial model $M(y_{..}, p_{ij})$ for the data y_{ij} in Table 3 the complete class lemma is again applied for Testing H_3 to give

$$\begin{array}{lcl} \text{the doubly cumulative chi-squared} & : & \chi^{**2} = \Sigma\Sigma\chi_{ij}^2, \\ \text{the maximal component of } \chi^{**2} & : & \max\max\chi_{ij}^2, \end{array}$$

where the χ_{ij}^2 are the goodness of fit chi-squared for the 2×2 tables obtained from the partitioning and accumulating of rows and columns at $i = 1, \cdots, a-1$ and $j = 1, \cdots, b-1$, respectively. The χ^{**2} is for the two-sided version of H_3 and $\max\max\chi^2$ applicable for both of the one and two-sided problems. When applied to Table 3 the two-sided p values are approximately 0.0065 for the χ^{**2} and exactly 0.0142 for $\max\max\chi^2$. The details of the p value caluculations and other variations of the test statistics are given in Hirotsu (1992, 97).

References

Armitage, P. (1955). Test for linear trend in proportions and frequencies. *Biometrics* **11**, 375-386.

Cochran, W. G. (1954). Some methods for strengthening the common χ^2 test. *Biometrics* **10**, 417-451.

Cohen, A. and Sackrowitz, H. B. (1991). Tests for independence in contingency tables with ordered categories. *Journal of Multivariate Analysis* **36**, 56-67.

Hawkins, D. M. (1977). Testing a sequence of observations for a shift in location. *J. Amer. Statist. Assoc.* **72**, 180-186.

Hirotsu, C. (1978). Ordered alternatives for interaction effects. *Biometrika* **65**, 561-570.

Hirotsu, C. (1979a). The cumulative chi-squares method and Studentized maximal contrast method for testing an ordered alternative in a one-way analysis of variance model. *Rep. Statist. Appl. Res., Union of Japanese Scientists and Engineers* **26**, 12-21.

Hirotsu, C. (1979b). An *F*-approximation and its application. *Biometrika* **66**, 577-584.

Hirotsu, C. (1982). Use of cumulative efficient scores for testing ordered alternatives in discrete models. *Biometrika* **69**, 567-577.

Hirotsu, C. (1986). Cumulative chi-squared statistic as a tool for testing goodness of fit. *Biometrika* **73**, 165-173.

Hirotsu, C. (1988). A class of estimable contrasts in an age-period-cohort model. *Ann. Inst. Statist. Math.* **40**, 3, 451-465.

Hirotsu, C. (1992). *Analysis of experimental data, beyond analysis of variance* (in Japanese). Tokyo:Kyoritsu-shuppan.

Hirotsu, C. (1997). Two-way change-point model and its application. *Austral. J. Statist.*(to appear).

Hirotsu, C., Kuriki, S. and Hayter, A. J. (1992). Multiple comparison procedure based on the maximal component of the cumulative chi-squared statistic. *Biometrika* **79**, 381-392.

Marcus, R. (1976). The powers of some tests of the equality of normal means against an ordered alternative. *Biometrika* **63**, 177-183.

Nair, V. N. (1986). On testing against ordered alternatives in analysis of variance model. *Biometrika* **73**, 493-499.

Simpson, D. G. and Margolin, B. H. (1986). Recursive nonparametric testing for dose response relationships subject to downturn at high doses. *Biometrika* **73**, 589-596.

Williams, D. A. (1971). A test for differences between treatment means when several dose levels are compared with a zero dose control. *Biometrics* **27**, 103-117.

Worsley, K. J. (1983). The power of likelihood ratio and cumulative sum tests of a change in a bionomial probability. *Biometrika* **70**, 455-464.

Worsley, K. J. (1986). Confidence regions and tests for a change point in a sequence of exponential family of random variables. *Biometrika* **73**, 91-104.

Tests for Linearity and Tests for Zero Slope in Crossover Studies

GÜNTHER HEIMANN

Schering AG, Corporate Biometry
13342 Berlin, Germany
e-mail: guenther.heimann@schering.de

Key Words:
crossover clinical trials, period effects, dose linearity, hypothesis testing, F-test

Abstract:
In a clinical phase II study the goal was to show the dose linearity and dose dependency of a test drug. Three different doses of the drug were tested in a group of patients. The study was conducted as a three period crossover experiment. The patients were randomized to the six different sequences of the test drug. The main target variables were pharmacokinetic parameters like AUC.

An appropriate statistical model to test departures from dose linearity is introduced. This model includes a sequence and a period effect. A very general theory (see Roebruck 1982, 1983) to construct optimal invariant F tests in mixed models is reviewed and explained in a geometrical setup. The methods are applied to the model as defined above. First, the hypotheses are stated in a precise way. Then the corresponding tests statistics are presented and it is shown how to analyze the corresponding data using SAS.

*Kitsos, C.P. and Edler, L., Eds.:Industrial Statistics, Physica-Verlag, Heidelberg, 1997

1. Introduction

This paper was motivated by a dose escalation study for an oral formulation of Fludarabine in patients suffering from low grade Non Hodgkins Lymphoma (lg NHL) and Chronic Lymphatic Leukemia (CLL). An intravenous (i.v.) formulation of Fludarabine is a well established medication for pretreated patients in these indications. Patients usually receive 6 to 8 treatment cycles (5 days of treatment followed by 23 days without treatment) of i.v. Fludarabine. In the present study, three doses of oral fludarabine (50 mg, 70 mg and 90 mg) were tested in 18 patients. Each patient received each of the doses for the first treatment day of the first three treatment cycles. The patients were randomly assigned to the six possible sequences (three patients to each sequence). The target variable was the area under the curve (AUC 24 h), which is a well established pharmacokinetic parameter to measure the exposure of patients with the test drug.

In this paper tests and hypotheses for linearity and tests for dose dependency in related crossover models will be presented. The focus is on the methodology, not on data analysis. By applying a general technique (Roebruck (1978), Roebruck (1982)) it is shown that these tests are best invariant tests. Also, it is indicated how to calculate these tests using SAS PROC GLM.

In the recent past, several authors have dealt with related problems. Patel and Tsai (1994) introduced a two-factor model with interactions and covariates, where one factor and the interaction are random, and considered tests about the covariance matrix. Güven (1995) considered a linear regression model with an extra random factor and maximum likelihood estimation of the parameters. Her model and other related models arise in calibration problems. Testing problems in these models are of interest for the calculation of detection and quantification limits (see DIN Norm 32645 for the non-mixed model case). Gallo and Khuri (1990) consider a two factor model with interactions where one factor and the interaction are assumed random, and Khuri and Littell (1987) consider the same model with all factors random. Both papers do not consider covariates. A more general approach was chosen by Seifert (1981), who deals with a very broad class of mixed models.

Related literature considers conditions for the equality of the Least Squares Estimator (LSE) and the Best Linear Unbiased Estimator (BLUE) in linear models with very general covariance structure. These conditions are related to the compatibility of the eigenspaces of the covariance matrix and the space generated by the design matrix, which is similar to the methodology used here and introduced by Roebruck (1978). Papers in this area are those of Zyskind (1967), Mc Elroy (1967), Milliken and Albohali (1984) and Baksalary and Eijnsbergen (1985). The list is by far not complete; the paper of Puntanen and Styan (1989) contains an extensive bibliography of this topic. None of the papers cited above treats the linearity testing problem, or the problem to test for the influence of covariates or dose. Before stating the model to be discussed in this paper, a short review is given on tests for linearity in the classical linear regression setting. The usual regression model has to be expanded according to

(1.1) $$Y_{ij} = \mu + \delta X_i + \gamma_i + \varepsilon_{ij}$$

where the parameters γ_i measure the deviance from linearity,. Defining

(1.2) $$Y = (Y_{11}, \ldots, Y_{1n_1}, \ldots, Y_{I1}, \ldots, Y_{In_I})^T$$

to be the column vector of observations, the vector ε in a similar way, and

(1.3) $$\vartheta = (\mu, \delta, \gamma_1, \ldots, \gamma_I)^T$$

to be the parameter vector, one can restate the model as

(1.4) $$Y = C\vartheta + \varepsilon$$

with

(1.5) $$C = [C_\mu \; C_\delta \; C_{\gamma_1} \ldots C_{\gamma_I}]$$

being the design matrix. The testing problem of linearity is then

(1.6) $$H_0 : \mathbf{E}Y \in \text{im}\left[C_\mu C_\delta\right] \text{ versus } H_1 : \mathbf{E}Y \in \text{im } C \setminus \text{im}\left[C_\mu C_\delta\right]$$

where **E** denotes expectation and im C is the space generated by the columns of C. The corresponding F-Test is

(1.7) $$F = \frac{\frac{1}{I-2}\sum_i n_i(\overline{Y}_{i.} - \hat{\mu} - \hat{\delta}(X_i - \overline{X}.))^2}{\frac{1}{n.-I}\sum_i\sum_j (Y_{ij} - \overline{Y}_{i.})^2}$$

where the numerator is the projection onto im C Θ im $[C_\mu C_\delta]$ and the denominator is the projection onto $R^{n.}$ Θ im C ; $\hat{\mu}$ and $\hat{\delta}$ are the usual least squares estimators for μ and δ.

The above test is well known and assumes $n_i > 1$, at least for some $i \leq I$.

2. The Statistical Model

In the sequel, the model

(2.1) $$Y_{ijk} = \mu + \alpha_i + \beta_j + \gamma_{l(i,j)} + \delta X_{l(i,j)} + U_{ik} + V_{ijk}$$

will be considered, where μ denotes the global mean; α ((i=1,...,I) the sequence effect; β_j (j=1,...,J) the period effect and δ the regression parameter. $X_{l(i,j)}$ defines the dose administered in sequence i and period j and $\gamma_{l(i,j)}$ defines the parameter to measure deviance from linearity.

U_{ik} (k=1,...,K_i) is the random patient effect; V_{ijk} is the random error. Carry over effects were not expected to be relevant here. In this notation,

(2.2) $$l(.,.) : \{1,\ldots,I\}\text{x}\{1,\ldots,I\} \rightarrow \{1,\ldots,L\}$$

may be viewed as a function, $L \geq 3$ is the total number of doses which may occur in the trial. Note that the number of periods J may be smaller than the number of doses L (i.e. the patients do not need to be treated with all doses that may occur in the trial) and that the number I of sequences equals

L!/(L-J)!. Please note also that not all possible I sequences need to occur in the trial since $K_i = 0$ is allowed for some $i \leq I$ in this paper.
As usually, the random variables $U_{11},\dots,U_{IK_I}$ will be assumed to be mutually independent and identically distributed, following a normal distribution with zero mean and variance σ_U^2. The assumption of mutual independence is also made for the random variables $V_{111},\dots,V_{IJK_I}$, which are normally distributed with zero mean and variance σ_V^2. Of course, the U_{ij} and V_{ijk} are independent of each other. Now define

(2.3)

$$Y = (\underbrace{\underbrace{Y_{111},\dots,Y_{1J1}}_{\text{1st patient}}\dots\underbrace{Y_{11K_1},\dots,Y_{1JK_1}}_{K_1\text{th patient}}}_{\text{1st sequence}},\dots,\underbrace{\underbrace{Y_{I11},\dots,Y_{IJ1}}_{\text{1st patient}}\dots\underbrace{Y_{I1K_I},\dots,Y_{IJK_I}}_{K_I\text{th patient}}}_{I-\text{th sequence}})^T$$

and

$$U = (\overbrace{\overbrace{U_{11},\dots,U_{11}}^{\text{1st patient}},\dots,\overbrace{U_{1K_1},\dots,U_{1K_1}}^{K_1\text{th patient}}}^{},\ \dots,\ \overbrace{\overbrace{U_{I1},\dots,U_{I1}}^{\text{1st patient}},\dots,\overbrace{U_{IK_I},\dots,U_{IK_I}}^{K_I\text{th patient}}}^{})^T$$

Let V be defined in the same way as Y with V_{ijk} as components instead of Y_{ijk}, and let

(2.4) $$\vartheta = (\mu,\alpha_1,\dots,\alpha_I,\beta_1,\dots,\beta_J,\delta,\gamma_1,\dots,\gamma_L)^T$$

With this notation one may represent the model (1) in matrix notation according to

(2.5) $$Y = C\vartheta + U + V$$

where

(2.6) $$C = \left[C_\mu C_{\alpha_1}\dots C_{\alpha_I} C_{\beta_1}\dots C_{\beta_J} C_\delta C_{\gamma_1}\dots C_{\gamma_L}\right]$$

is the appropriate design matrix with column C_μ corresponding to μ, C_{α_1} to α_1, etc. All columns consist of zeroes and ones and are defined in an obvious way. Since

(2.7) $$C_\mu = \sum_{i=1}^{I} C_{\alpha_i} = \sum_{j=1}^{J} C_{\beta_j} = \sum_{l-1}^{L} C_{\gamma_l}$$

and

(2.8) $$C_\delta = \sum_{l=1}^{L} X_l C_{\gamma_l}$$

one easily obtains

(2.9) $$\mathrm{rank}(C) = 1 + (I-1) + (J-1) + (L-1).$$

The testing problems that we are mainly interested in here refer to the parameters $\gamma_1,\ldots,\gamma_L$ for the test of linearity and to the parameter δ for the test of dose dependency. The testing problems will be specified in detail in the next sections.

3. The F-Test for Linearity

To be mathematically precise, the hypothesis of linearity may be expressed in terms of a linear subspace of the IK. - dimensional real space, namely

(3.1) $$H_0 : \mathbf{E}Y \in \mathrm{im}\, H := \mathrm{im}\left[C_\mu C_\alpha \ldots C_{\alpha_I} C_{\beta_1} \ldots C_{\beta_J} C_\delta\right]$$

where "**E**" denotes the expected value of the respective random vector, im(A) denotes the space spanned by the columns of a matrix A, and where

$$K. = \sum_{i=1}^{I} K_i$$

The hypothesis is equivalent to

(3.2) $$H_0 : \mathbf{E}Y \text{ is orthogonal to im (C)} \ominus \text{im (H)}$$

where im(C)$\ominus$ im(H) is the orthogonal complement of im(H) in im(C). The alternative hypothesis is

(3.3) H_1 : **E**Y is not orthogonal to im (C) Θ im (H)

The dimension of im(C) equals (I-1)+(J-1)+(L-1)+1, and the dimension of im(H) equals I+J. The null hypothesis may be stated in terms of the parameter vector as linear combinations that only involve the parameters $\gamma_1,\ldots,\gamma_L$.

The corresponding F-test is based upon

(3.4)

$$F = \frac{(K.(J-1)-L-J+2)\, Y^T(C(C^TC)^-C^T - H(H^TH)^-H^T)Y}{(L-2)Y^T(E_{JK.} - E_1(E_1^TE_1)^{-1}E_1^T - C(C^TC)^-C^T + C_\alpha(C_\alpha^TC_\alpha)^{-1}C_\alpha^T)Y}$$

where E_1 and C_α are block diagonal matrices

(3.5)

$$E_1 = \mathrm{diag}(1_J \ldots 1_J)$$

$$C_\alpha = [C_{\alpha_1} \ldots C_{\alpha_1}]$$

of dimensions (JK.)x(K.) and (JK.)x(I). $E_{JK.}$ is the unit matrix of corresponding dimension and 1_J is a column vector of length J with all entries equal to 1. A^{-1} and A^- denote the inverse and a generalized inverse of a matrix A.

In case of equal numbers of patients per sequence, the test statistic may be calculated as

(3.6)
$$F = \frac{(KI-2)(J-1)IK\sum_{l=1}^{L}(\overline{Y}_{(l)} - \hat{\mu} - \hat{\delta}(X_l - \overline{X}.))^2}{(L-2)\sum_{i=1}^{I}\sum_{k=1}^{K}\sum_{j=1}^{J}(Y_{ijk} - \overline{Y}_{i.k} - (\overline{Y}_{.j.} - \overline{Y}\ldots) - (\overline{Y}_{(l(i.j))} - \overline{Y}\ldots))^2}$$

where $\overline{X}., \overline{Y}_{.j.}, \overline{Y}\ldots$ etc. denote the respective mean values and where $\overline{Y}_{(l)}$ or $\overline{Y}_{(l(i,j))}$ denote the mean value of all observations associated with dose l , l(i,,j) respectively.

$\hat{\mu}$ and $\hat{\delta}$ are the usual estimators of slope and intercept.

In case of J=L, i.e. if every patient receives every dose, the sequence effect may as well be omitted, and the model

$$(3.7) \qquad Y_{ijk} = \mu + \beta_j + \gamma_{l(i,j)} + \delta X_{l(i,j)} + U_{ik} + V_{ijk}$$

may be chosen instead of (2.1). (For the notation see section 2). Choosing $L = \left[C_\mu C_\beta \ldots C_{\beta_l} C_\delta\right]$, the hypothesis of linearity may be stated as above, and it turns out that the corresponding F-test is again based upon (5). One interpretation of the sequence effect in this model is therefore that it describes the effect of the total dose that a patient received.

In order to calculate the test using SAS PROC GLM, one needs a dataset DATA with five columns Y, A, B, D and U. The column Y contains the target variable, A contains the information upon the sequence, B the information upon the period, and D the actual dose level. U contains the patient number. The basic SAS statement is then

```
C = D;
PROC GLM DATA = DATA,
CLASS A B C U;
MODEL Y = A U(A) B D C ;
RANDOM U(A);
```

Note that the random patient effect U has to be nested within A in order to obtain the correct tests, and that the ordering of the factors in the model statement is important.

4. The F-Test for Dose Dependency

The hypotheses of "dose dependency" may be stated in a precise way as

$$(4.1) \qquad H_0 : \text{EY is orthogonal to im } H \ominus \text{im } G$$

where $G = \left[C_\mu C_{\alpha_1} \ldots C_{\alpha_1} C_{\beta_1} \ldots c_{\beta_J}\right]$. The alternative hypothesis is

(4.2) $$H_1 : \mathbf{E}Y \text{ is not orthogonal to im } H \ominus \text{im } G,$$

and the dimension of im G equals I+J-1. Note that (1) is equivalent to

(4.3) $$H_0 : IK\sum_l (X_l - \overline{X}.)^2\delta + IK\sum_l (X_l - \overline{X}.)\gamma_l = 0$$

The corresponding F-Test is

(4.4)

$$F = \frac{(K.(J-1)-J-L+2)Y^T(H(H^TH)^-H^T - G(G^TG)^-G^T)Y}{Y^T(E_{JK.} - E_l(E_l^TE_l)^{-1}E_l^T - C(C^TC)^-C^T + C_\alpha(C_\alpha^TC_\alpha)^{-1}C_\alpha^T)Y}$$

(For the notation, see section 3.) Again, in case of equal numbers of patients per sequence the test statistic may be presented as

(4.5) $$F = \frac{(IK-2)(J-1)\, IK \sum_{l=1}^{L}(X_l - \overline{X}.)^2\hat{\delta}^2}{\sum_{i=1}^{I}\sum_{k=1}^{K}\sum_{j=1}^{J}(Y_{ijk} - \overline{Y}_{i.k} - (\overline{Y}_{.j.} - \overline{Y}\ldots) - (\overline{Y}_{(l(i.j))} - \overline{Y}\ldots))^2}$$

In the case of J=L one may again drop the sequence effect from the model and work with (3.8). The resulting F-test is then still the one based upon (4), though, so the inclusion or exclusion of the sequence effect has no impact. The SAS statement is the same as in section 3. It is important to use type I sums of squares here. There are no type II and type III sums of squares associated with δ in this model H. Precisely: the respective type II and type III sums of squares are equal to zero.

If one is interested in a test for $H_0{:}\delta = 0$, then one needs to look at an other model, namely

(4.6) $$Y = H\vartheta + U + V$$

cf. (2.6) and (3.1). The respective test statistics would have the same numerator as (4) and (5), but a different denominator;

$$(4.7)\quad F = \frac{(K.(J-1)-J)Y^T(H(H^TH)^-H^T - G(G^TG)^-G^T)Y}{Y^T(E_{JK.} - E_1(E_1^TE_1)^{-1}E_1 - H(H^TH)^-H^T + C_\alpha(C_\alpha^TC_\alpha)^{-1}C_\alpha^T)Y}$$

5. Derivation and Optimality of the Tests

In this section it will be briefly sketched how to derive the two test statistics (3.5) and (4.4), and how to use the results of Roebruck (1978), (1982) and (1983) to prove their optimality. The covariance matrix of the random vector Y defined in (2.4) is

$$(5.1)\qquad \Lambda = \begin{bmatrix} \Psi & . & . & . & O_{JxJ} \\ . & . & & & . \\ . & & . & & . \\ . & & & . & . \\ O_{JxJ} & . & . & . & \Psi \end{bmatrix} = diag(\Psi \ldots \Psi)$$

which is of dimension ((JK.)x(JK.)). The JxJ matrix Ψ is defined as

$$(5.2)\qquad \Psi = \sigma_v^2 E_J + \sigma_u^2 1_{JxJ}$$

0_{JxJ} and 1_{JxJ} are matrices of dimension JxJ where all entries are zeroes or ones, respectively: One can easily see that (5.1) has two eigenvalues

$$(5.3)\qquad \lambda_1 = \sigma_v^2 + J\sigma_u^2 \quad , \quad \lambda_2 = \sigma_v^2$$

with multiplicities

$$(5.4)\qquad \mu_1 = K. \;\; , \;\; \mu_2 = K.(J-1)$$

The eigenvectors of λ_1 are the columns of the matrix

$$(5.5)\qquad E_1 = diag(1_J \ldots 1_J)$$

which is of dimension JK. x K., cf. (3.6). Let

$$(5.6)\qquad P_A = A(A^TA)^-A = P_{im\,(A)}$$

denote the projection operator onto the subspace im(A) for any matrix A. One can immediately derive

(5.7) $$P_{E_1} C_\beta = P_{C_\alpha} C_\beta \quad \text{for } C_\beta = \left[C_{\beta_1} \ldots C_{\beta_J}\right]$$

and

(5.8) $$P_{E_1} C_\gamma = P_{C_\alpha} C_\gamma \quad \text{for } C_\gamma = \left[C_{\gamma_1} \ldots C\gamma_L\right]$$

as well as

(5.9) $$P_{E_1} C_\delta = P_{C_\alpha} C_\delta$$

From this one obtains

(5.10) $$\text{im } C_\alpha \subseteq E(\lambda_1)$$

where $E(\lambda_i)$ are the eigenspaces,

(5.11) $$\begin{aligned} im\ C \ominus \text{im } C_\alpha &= \text{im}\,(E_{JK} - P_{C_\alpha})\left[C_\beta C_\delta C_\gamma\right] \\ &= \text{im}\,(E_{JK} - P_{E_1}\left[C_\beta C_\delta C\gamma\right] \subseteq E(\lambda_2) \end{aligned}$$

and im C $\ominus$ im C_α is the orthogonal complement of im $(C_{\alpha)}$ within im (C). Now, from (10) and (11) one concludes that

(5.12) $$\text{im (C)} \ominus \text{im (H)} \subseteq \text{im (C)} \ominus \text{im } (C_\alpha) \subseteq E(\lambda_2)$$

and since

(5.13) $$P_{\text{im C} \ominus \text{im H}} = C(C^T C)^- C^T - H(H^T H)^- H^T$$

one may further conclude that the numerator of (3.5) divided by λ_2 follows a chi square distribution with L-2 degrees of freedom, which is central under H_0.

The space $E(\lambda_2) \ominus$ (im C $\ominus$ im C_α) is than orthogonal to im C $\ominus$ im H, and the projection operator onto this space is

(5.14) $E_{JK.} - E_1(E_1^T E_1)^{-1} E_1 - C(C^T C)^- C^T + C_\alpha (C_\alpha^T C_\alpha)^{-1} C_\alpha^T$

Therefore, the denominator of the test statistics divided by λ_2 follows a central chi square distribution with K.(J-1)-J-L+2 degrees of freedom, and is stochastically independent from the numerator of (3.5). This shows that the test statistic (3.5) follows an appropriate central F distribution under H_0. Similar arguments, of course, yield (4.4).

The optimality of these tests may be verified using the methodology of Roebruck (1978) (see also Roebruck (1982), and Roebruck (1983)). The model and hypothesis are regular in the sense of def. 2.10 and 3.1 in Roebruck (1978). Basically, this means that there is a decomposition of the spaces that are relevant for the testing problems into subspaces of the eigenspaces of Λ. In the current model for the testing problem (3.1) this means that (n=JK.!).

$$\text{im } \mathbf{R}^n \ominus \text{im } C = E(\lambda_1, \text{im } \mathbf{R}^n \ominus \text{im } C) \oplus E(\lambda_2, \text{im } \mathbf{R}^n \ominus \text{im } C)$$

$$\text{im } C \ominus \text{im } H = E(\lambda_2, \text{im } C \ominus \text{im } H)$$

$$\text{im } H = E(\lambda_1, \text{im } H) \oplus (\lambda_2, \text{im } H)$$

and that

$$E(\lambda_1) = E(\lambda_1, imH) \oplus \varepsilon E(\lambda_1, im\mathbf{R}' \ominus imC)$$
$$E(\lambda_2) = E(\lambda_2, imH) \oplus \varepsilon(\lambda_2, imC \ominus imH) \oplus E(\lambda_2, im\mathbf{R}'' \ominus imC)$$

The notation $E(\lambda_i, M)$ means that $E(\lambda_i, M)$ is both a subspace of $E(\lambda_i)$ as of M.

In chapter 3 of Roebruck (1978) it is shown that there is a suitable group (see 3.6, Roebruck (1978))

$$G = \left\{Ax + d: \in \text{im H}; AA^T = E_{JK.}, AP_{\text{im C}} = P_{\text{im C}}, AP_{\text{im H}} = P_{\text{im H}}, AP_{E(\lambda i)} = P_{E(\lambda i)}\right\}$$

of transformations $Ax + d$ from JK. - dimensional space onto itself, and that the test based upon (3.5) is the best invariant test with respect to this group G

The above techniques prove to be useful in many crossover models including those where one wants to model carryover effects and where one is interested in test on sequence, period, treatment or carryover effects, or in crossover models where one wants to model a covariate.

Bibliography

Baksalary J.K. and Van Eijnsbergen A.C. (1988)
A Comparison of Two Criteria for Ordinary Least Squares Estimators to Best Linear Unbiased Estimators. The American Statistician 42, 205-208

Gallo, J. and Khuri A.I. (1990)
Exact Tests for the Random and Fixed Effects in an Unbalanced Mixed Two-Way Cross Classification Model. Biometrics 46, 1087-1095

Güven B. (1995)
Maximum Likelihood Estimation in Simple Linear Regression With One-Fold Nested Error. Communications in Statistics - Theory & Methods 24(1), 121-130

Khuri A.I. and Littell R.C.(1987)
Exact Tests for the Main Effects Variance Components in an Unbalanced Random Two-way Model. Biometrics 43, 545-560

Mc Elroy, F.W. (1967)
A Necessary and Sufficient Condition That Ordinary Least Squares Estimators are Best Linear Unbiased. JASA 62, 1302-1304

Milliken G.A. and Albohali M. (1984)
On Necessary and Sufficient Conditions for Ordinary Least Squares to Be Best Linear Unbiased Estimators. The American Statistician 38, 298-299

Patel H.I. and Tsai K.T. (1994)
Tests of Special Sturctures on the Dispersion Matrix Associated with a Two Way Mixed Effects Model. Biometrical Journal 36-8,953-961

Puntanen S. and Styan G.P.H. (1989)
The Equality of the Ordinary Least Squares Estimator and the Best Linear Unbiased Estimator. The American Statistician 43, 153-161

Roebruck, P. (1978)
Über optimale Tests in linearen Modellen mit gemischten Effekten.
Dissertation, Technische Hochschule Aachen

Roebruck, P. (1982)
On the Optimality of F-test on Variance Components in Mixed Linear Models.
Biometrical Journal 24-8, 811-815

Roebruck, P. (1983)
Zur Optinalität des F-Tests für lineare Hypothesen über lineare Modelle mit gemischten Effekten. Metrika 30, 85-91

Seifert, B.(1981)
Explicite Formulae of Exact Tests in Mixed Balanced ANOVA-models.
Biometrical Journal 23-6, 535-550

Part V

Repeated Measurements - Multiple Inference

A Nonparametric Combination Method For Dependent Permutation Tests with Application to Some Problems with Repeated Measures

FORTUNATO PESARIN
Department of Statistics, University of Padova
Via San Francesco, 33, 35121 Padova, Italy

Key words and phrases: Dependent random effects, longitudinal data, nonparametric test combination, permutation tests, repeated measures, stochastic additive effects, stochastic dominance.

Abstract

The paper deals with a nonparametric combination method for dependent permutation tests for the solution of complex multivariate testing problems. The method is based on: a) a decomposition of the hypotheses into $k > 1$ sub-hypotheses, where for each sub-hypothesis an unbiased, consistent and significant for large values permutation partial test is supposed to exist; b) a simulation procedure, conditional on the observed data, to estimate the null multivariate permutation distribution of partial tests; c) a combination of partial tests into a second-order statistic, whose null permutation distribution is estimated by using the same simulation results of first step. As a motivating example, it is also discussed an application to some testing problems on treatment effects when measurements are repeated on the same units at fixed times and with dependent errors. Several Monte Carlo results show the behaviour of the given solution for a stochastic dominance problem.

1 Introduction

This paper deals with a general testing problem on treatment effects when measurements are repeated on the same units (or individuals) at fixed time

*Kitsos, C.P. and Edler, L., Eds.:Industrial Statistics, Physica-Verlag, Heidelberg, 1997

occasions. Units are supposed to be partitioned into C groups or samples, according to C levels of a symbolic treatment and the hypotheses under testing are related to whether the observed profiles of repeated measures are or are not dependent on treatment levels. This problem is discussed both to motivate the effectiveness of the *method of nonparametric combination of several dependent permutation tests* and to give a solution to rather difficult problems. In this paper, repeated measures, panel data, longitudinal data, response trajectories and profiles are considered as synonymous. The proposed solutions are essentially based on a combination of dependent permutation tests, each obtained by means of a *time by time analysis* (PESARIN, 1992; PALLINI & PESARIN, 1995). To this end, we will assume the validity of the permutation testing principle, i.e. under the null hypothesis, the individual profiles are assumed exchangeable with respect to treatment levels. It is useful to underline that permutation tests are conditional on the whole set of observed data which is always a set of sufficient statistics for whichever underlying model. Hence, permutation tests are data dependent distribution-free procedures. This implies that no assumption on distributions of error terms is necessary, apart that of time by time homoscedasticity.

Measures of a univariate variable Y are supposed to be independent with respect to units, but not with respect to time. To be more specific, let us refer to a problem with n units partitioned into C groups. Groups are of size $n_j \geq 2$, $j = 1, ..., C$, with $n = \sum_j n_j$. Individuals belonging to j-th group are supposed to receive a symbolic treatment at j-th level. All units are observed at k fixed occasions $\tau_1, ..., \tau_k$, where k is a finite integer. For simplicity we refer to time occasions by using t to mean τ_t, $t = 1, ..., k$. Hence, for each unit we suppose to observe a profile of a stochastic process and profiles related with different units are assumed independent. A profile can be viewed either as a discretized realization of an underlying stochastic process or as a k-dimensional random variable. We assume that there are no missing values. Moreover, we only refer to a unidimensional stochastic model with additive effects covering a number of practical situations.

2 A model with additive effects

The symbol $\mathbf{Y} = \{Y_{ji}(t),\ i = 1, ..., n_j,\ j = 1, ..., C,\ t = 1, ..., k\}$ is used to indicate that the whole set of observed data is organized in a Two Way layout. Alternatively, $\mathbf{Y}$ can be organized as a One Way layout of profiles: $\mathbf{Y} = \{\mathbf{Y}_{ji}, i = 1, ..., n_j, j = 1, ..., C\}$, where $\mathbf{Y}_{ji} = \{Y_{ji}(t), t = 1, ..., k\}$ indicates the ji-th observed profile.

The response model we are referring to is

$$Y_{ji}(t) = \mu + \eta_j(t) + \delta_{ji}(t) + \sigma(t) Z_{ji}(t), i = 1, ..., n_j, j = 1, ..., C, t = 1, ..., k, \quad (1)$$

where: $Z_{ji}(t)$ are error terms distributed as a stationary stochastic process with null mean and with distribution P_Z, not necessarily Gaussian; the $Z_{ji}(t)$

are assumed to be i.i.d. with respect to units and treatment levels but not independent on time; μ is a population constant; coefficients $\eta_j(t)$ represent treatment effects and may depend on time; coefficients $\delta_{ji}(t)$ represent individual effects. When $\delta_{ji}(t)$ are stochastic, we assume that they are with null mean values and variances not dependent on units and on treatment levels. Hence, the random effects $\delta_{ji}(t)$ are realizations of a k-variate variable Δ with unspecified distribution $\mathcal{D}_k\{\mathbf{0}, \beta\}$, where the matrix β, indicating how units vary between themselves, is not dependent on treatment levels. This setting is consistent with a general form of dependent random effects. Of course, when $\beta = \mathbf{0}$ for all t, the model is with fixed effects. Time-varying factors $\sigma^2(t)$ are assumed to be independent from units and treatment levels. In order to appreciate how difficult the statistical analysis of real problems is, see AZZALINI (1984), CROWDER & HAND (1990), DIGGLE, LIANG & ZEGER (1994). In particular when the matrix β is without any known structure, the model is not identifiable and thus no parametric inference is possible.

One specification of this dependent random effects model assumes that coefficients $\delta_{ji}(t)$ behave according to an AR(1) model, i.e. such that $\delta_{ji}(0) = 0$ and

$$\delta_{ji}(t) = \nu(t)\cdot\delta_{ji}(t-1) + \beta(t)\cdot W_{ji}(t), i = 1, ..., n_j, j = 1, ..., C, t = 1, ..., k \quad (2)$$

where: $W_{ji}(t)$ represent random i.i.d. innovations on coefficients Δ and interpret the deviates of individual behaviour; $\nu(t)$ represent autoregressive coefficients which are assumed independent on treatment levels and on units, but not on time; $\beta(t)$, $t = 1, ..., k$, are time varying coefficients of variability of AR coefficients Δ. By assumption the innovations $W_{ji}(t)$ have null mean value. This setting is consistent with a form of autocorrelated random effects.

3 The hypotheses of interest

The hypotheses that we wish to test are:

$$\begin{aligned} H_0 &= \{\mathbf{Y}_1 \overset{d}{=} \ldots \overset{d}{=} \mathbf{Y}_C\} = \{Y_1(t) \overset{d}{=} \ldots \overset{d}{=} Y_C(t), t = 1, \ldots, k\} = \\ &= \bigcap_{1\le t\le k} \{Y_1(t) \overset{d}{=} \ldots \overset{d}{=} Y_C(t)\} = \bigcap_{1\le t\le k} H_{0t}, \end{aligned} \quad (3)$$

against $H_1 = \{H_0 \text{ is not true}\} = \bigcup_{t=1}^{k}\{H_{0t} \text{ is not true}\} = \bigcup_{t=1}^{k} H_{1t}$, where a suitable *time by time decomposition* of the hypotheses into k sub-hypotheses is highlighted.

The form $H_0 = \{\mathbf{Y}_1 \overset{d}{=} \ldots \overset{d}{=} \mathbf{Y}_C\}$ implies that profiles are exchangeable with respect to treatment levels. This situation is standard for most experimental designs when units are randomly assigned to treatment levels and are assumed to be homogeneous with respect to most important experimental conditions, such as age, sex, health, etc.. Thus, the permutation testing principle directly applies.

Assuming homoscedasticity of profiles, the hypotheses may be specified as follows

$$H_0 = \{\eta_1(t) = ... = \eta_C(t), t = 1, ..., k\} = \bigcap_{1 \leq t \leq k} \{\eta_1(t) = ... = \eta_C(t)\}, \quad (4)$$

against $H_1 = \{H_0 \text{ is not true}\} = \bigcup_{t=1}^{k} H_{1t}$.

Note that each of the k sub-problems is reduced to a One Way ANOVA. Hence, for testing separately H_{0t} against H_{1t} we may use a suitable test, T_t say. For quantitative univariate data, such tests have the standard C-sample Snedecor's type form, because homoscedasticity of profiles implies homoscedasticity of data for every fixed t. It should be noted that each partial Snedecor's type test is marginally unbiased and consistent and that large values are significant.

Suppose the k partial tests T_t, $t = 1, ..., k$, are available. In order to achieve a complete solution for H_0 against H_1 we must combine all these partial tests through a suitable combining function. Of course, due to the complexity of the problem and to the unknown k-dimensional distribution of $T_1, ..., T_k$ (DIGGLE ET AL. 1994), we are unable to estimate all dependence coefficients among partial tests directly from the data $\mathbf{Y}$. Therefore, the combination should be nonparametric, especially with respect to these underlying unknown dependencies. It is worth noting that when the model is not identifiable this nonparametric combination becomes unavoidable.

When baseline observations $Y_{ji}(0)$ are present and are not identically distributed or when their variability is not negligible, then all partial tests should be conditional on them, i.e. $Y_{ji}(0)$ must be considered as covariates for the problem. In practice the testing can be done by taking into consideration k ANCOVA layouts. Sometimes, it is sufficient to take into consideration time by time increments: $Y_{ji}(t) - Y_{ji}(t-1)$, or even increments with respect to baseline: $Y_{ji}(t) - Y_{ji}(0)$. Of course, in these cases the k partial tests should be accordingly modified.

A rather special and interesting problem holds when $C = 2$ and when, for instance, we are interested in testing if the first process is stochastically dominated by the second (first order dominance). In such a case and referring to models with stochastic coefficients, the hypotheses are:

$$H_0 = \{Y_1(t) \overset{d}{=} Y_2(t), t = 1, ..., k\} = \bigcap_{1 \leq t \leq k} [\eta_1(t) = \eta_2(t)] = \bigcap_{1 \leq t \leq k} H_{0t}, \quad (5)$$

against $H_1 = \bigcup_{t=1}^{k} (Y_1(t) \overset{d}{<} Y_2(t)) = \bigcup_{t=1}^{k} [\eta_1(t) < \eta_2(t)] = \bigcup_{t=1}^{k} H_{1t}$, where a stochastic dominance problem is highlighted. Observe that the alternative is now decomposed into directional (or restricted) sub-alternatives. Hence, for each sub-hypothesis a one tailed Student's type test should be considered.

4 The nonparametric combination method

Without loss of generality, hereafter we assume all univariate permutation partial tests T_t of interest are marginally unbiased and significant for large values; i.e. the permutation marginal distribution of each T_t is stochastically larger under H_{1t} than under H_{0t}, $t = 1, ..., k$. The marginal unbiasedness and the significance for large values of partial tests imply that they are Positively Upper Orthant Dependent (PUOD). It is worth noting that conditioning on sufficient statistics provides permutation methods with nice inferential properties, especially with regard to their asymptotic behaviour. Unfortunately the cardinality of the permutation space $(\mathcal{Y}|\mathbf{Y})$ is generally too large to deal with. Hence, for a reasonable estimate of the k-dimensional null permutation distribution $F(\mathbf{x}|\mathbf{Y})$ of vector test $\mathbf{T}$, $\forall \mathbf{x} \in \mathcal{R}^k$, we suggest to use a form of *conditional simulation procedure.* The nonparametric combination method is fully discussed in PESARIN (1992) and in PALLINI & PESARIN (1995).

Let us indicate by $\mathbf{Y}$ both the set of observed data and the pooled sample $\mathbf{Y} = \mathbf{Y}_1 \oplus \mathbf{Y}_2 \oplus \ldots \oplus \mathbf{Y}_C$, i.e. a file with n rows, where: the first n_1 rows contain the profiles of units belonging to the first group; next n_2 rows, from the $n_1 + 1$ until the $n_1 + n_2$, contain the profiles belonging to the second group; and so on. We may denote this pooled sample by $\mathbf{Y} = \{Y(t;i),\ t = 1, ..., k,\ i = 1, ..., n,\ n_1, ..., n_C\}$.

In practice the nonparametric combination takes into consideration the permutation p-values λ_t associated with partial tests T_t, $t = 1, ..., k$. Thus, the combined test $T'' = \psi(\lambda_1, ..., \lambda_k)$ is achieved by a continuous, non-increasing, univariate and non-degenerate real function $\psi \quad (0,1)^k \to \mathcal{R}^1$. All combining functions ψ must satisfy the following conditions:

I) ψ must be non-increasing in each argument, i.e. $\psi(.., \lambda_t, ..) \geq \psi(.., \lambda'_t, ..)$ if $\lambda_t < \lambda'_t$, $\forall t \in (1, ..., k)$;

II) ψ must attain its supremum value $\overline{\psi}$, which could be not finite, when at least one argument attains zero, i.e. $\psi(.., \lambda_t, ..) \to \overline{\psi}$ if $\lambda_t \to 0$, $1 \leq t \leq k$;

III) its critical value T''_α is finite, i.e. it is such that: $T''_\alpha < \overline{\psi}$, $\forall \alpha > 0$.

These properties define a class $\mathcal{C}$ of combining functions, which contains the well known algorithms of Fisher, Lancaster, Liptak and Tippett (for a review of combination algorithms, see FOLKS, 1984). The class $\mathcal{C}$ also contains the Mahalanobis quadratic form and the Birnbaum complete class of admissible combining functions of independent tests. In particular $\mathcal{C}$ contains all combining functions which take care nonparametrically of the underlying dependence structure among the p-values λ_t, $t = 1, ..., k$.

In short, the conditional simulation procedure considers the following steps:

a) Calculate the observed value of T_t: $T_{obt} = T_t(Y)$, $t = 1, ..., k$.

b) i)Consider a random permutation $\{u_1, ..., u_n\}$ of integers $\{1, ..., n\}$; ii) assign first n_1 corresponding profiles to the first group, next n_2 to the second, and so on, obtaining a data permutation $Y^* = \{Y(t; u_i),\ t = 1, ..., k,\ i = 1, ..., n,\ n_1, ..., n_C\}$; iii) calculate test statistics $T_t^* = T_t(\mathbf{Y}^*)$, $t = 1, ..., k$.

c) Independently repeat B times step b). The set of resampling results $\{\mathbf{Y}_r^*,\ r = 1, ..., B\}$ is thus a Monte Carlo simulation of the null (permutation) multivariate distribution $F(\mathbf{x}|\mathbf{Y})$ of tests $\mathbf{T}$.

d) The multivariate E.D.F. $\widehat{F}_B(\mathbf{x}) = [1/2 + \sum_{r=1}^{B} \mathbf{I}(\mathbf{T}_r^* \leq \mathbf{x})]/(B+1)$, $\mathbf{x} \in R^k$, where $\mathbf{I}(\cdot)$ equals 1 if relation $(\cdot)$ is satisfied and 0 otherwise, by the well known Glivenko-Cantelli Theorem provides a consistent estimate of the distribution $F(\mathbf{x}|\mathbf{Y})$ of $\mathbf{T}$; moreover:

$$\widehat{L}_t(x) = [1/2 + \sum\nolimits_{1 \leq r \leq B} \mathbf{I}(T_{tr}^* \geq x)] \,/\, (B+1), x \in R^1,$$

gives a consistent estimate of the marginal permutation significance level functions $L_t(x) = Pr(T_t^* \geq x|\mathbf{Y})$ of T_t, $t = 1, ..., k$; thus, $\widehat{L}_t(T_{obt}) = \widehat{\lambda}_t$ gives a consistent estimate of the marginal p-value $\lambda_t = Pr(T_t^* \geq T_{obt}|\mathbf{Y})$ (these consistency properties are for B going to infinity).

e) The observed value of the combined test, still evaluated through the same conditional simulation results, is given by $T''_{ob} = \psi(\widehat{\lambda}_1, ..., \widehat{\lambda}_k)$.

f) The r-th combined test is calculated according to $T_r''^* = \psi(\lambda_{1r}^*, ..., \lambda_{kr}^*)$, where $\lambda_{tr}^* = \widehat{L}_t(T_{tr}^*)$, $t = 1, ..., k$, $r = 1, ..., B$.

g) Hence, the p-value of combined test T'' is consistently estimated by:

$$\widehat{\lambda}'' = \sum\nolimits_{1 \leq r \leq B} \mathbf{I}(T_r''^* \geq T''_{ob})/B.$$

h) If $\widehat{\lambda}'' < \alpha$, the general null hypothesis H_0 is rejected at significance level.

The most common combining algorithm is that of Fisher: $T''_F = -2 \cdot \sum_t log(\lambda_t)$. In this paper, provided that all partial tests are expressed in standardized form and that their asymptotic support is unbounded on the right, we will use the so called *direct* combination algorithm: $T''_D = \sum_t T_t = \sum_t L_T^{-1}(\lambda_t) = \sum_t L_T^{-1} \circ L_T(T_t)$, where L_T represents the common permutation significance level function of partial tests. The direct combination algorithm can be properly used when at least approximately all partial tests share the same permutation distribution, i.e. for quite large sample sizes. With respect to others, the direct combination algorithm is much less computer intensive, because step d) can be bypassed.

If all partial tests T_t are marginally unbiased and consistent for testing H_{0t} against H_{1t}, $t = 1, ..., k$, then the combined test T'' is unbiased and consistent for testing H_0 against H_1. In order to prove these properties, let us assume: 1) when n diverges, all the n_j, $j = 1, ..., C$, diverge; 2) the number B diverges; 3) k and α are fixed.

Let us first consider the consistency property:

If $\mathbf{T} = \{T_i, i = 1, ..., k\}$ *are consistent first order permutation tests, respectively for* H_{0i} *against* H_{1i}*, then* $\forall \psi \in \mathcal{C}$, $T'' = \psi(\lambda_1, ..., \lambda_k)$*, is a consistent combined test for* H_0 *against* H_1.

Arguments for the proof are: a) to be consistent a combined test must reach its critical region with probability one if at least one among H_{1i}, $i = 1, ..., k$, is true; b) by assumed consistency of $\mathbf{T}$, H_{1i} true implies $\lambda_i \to 0$ in probability as $n \to \infty$; c) by properties II) and III) of this section, $T'' \to \bar{\psi} > T''_\alpha$ with probability one, $\forall \alpha > 0$.

To state unbiasedness of T'', we do use the Proposition: *let* X *and* Z *be two variables defined on the same univariate probability space, if* X *is stochastically larger than* Z*, i.e. if* $F_X(x) \leq F_Z(x)$, $\forall x \in \mathcal{R}^1$*, and if* φ *is a non-decreasing real function, then* $\varphi(Y)$ *is stochastically larger than* $\varphi(Z)$, whose proof is straightforward. Hence:

If $\mathbf{T} = \{T_i, i = 1, ..., k\}$ *are unbiased first-order permutation tests respectively for* H_0 *against* H_{1i}*, i.e. the* T_i, $i = 1, ..., k$*, are PUOD, then* $\forall \psi \in \mathcal{C}$, $T'' = \psi(\lambda_1, ..., \lambda_k)$ *is an unbiased combined test for* H_0 *against* H_1.

Arguments for the proof are: a) unbiasedness and PUOD properties of first-order tests T_i imply $\Pr\{\lambda_i \geq z | H_{0i}\} \geq \Pr\{\lambda_i \geq z | H_{1i}\}$, $i = 1, ..., k$ and $\forall z \in (0, 1)$, i.e. the $(\lambda_i | H_{1i})$ are stochastically smaller than $(\lambda_i | H_{0i})$; b) by the non-decreasing property of ψ and Proposition above, $\psi(..., \lambda_i, ... | H_{1i})$ is stochastically larger than $\psi(..., \lambda_i, ... | H_{0i})$; c) hence, by iterating for $i = 1$ up to k, unbiasedness of T'' is achieved.

It should be noted that all combining functions based on convex linear functions of bounded scores are not consistent. In fact they violate property II) because they do not attain their critical region with probability one if only some of the sub-alternatives are true. So, a combined test of the form $\sum_i w_i \cdot \phi(\lambda_i)$, where the weights are such that $w_i > 0$, $\sum_i w_i = 1$ and the $\phi(\lambda_i)$ are *positive bounded scores*, is not consistent. Hence it is not a member of $\mathcal{C}$. Moreover, in Pallini (1994) it is proved that if all partial tests are Bahadur efficient the combining algorithms of Fisher, Lancaster and Liptak are Bahadur efficient too.

5 Solutions of testing problems

Testing solutions are now straightforward. To this end we observe that: a) the permutation principle applies, because the exchangeability of profiles with respect to treatment levels is satisfied under H_0; b) according to (3), (4), and

(5), the hypotheses are suitably decomposed into a set of sub-hypotheses; c) by assumption, for each sub-hypothesis there exists a test which is marginally unbiased, consistent and significant for large values. Hence, the nonparametric combination method applies and solutions are at least unbiased and consistent.

To be specific, when data are according to model (1), or when the baseline observations $Y_{ji}(0)$, if present, are i.i.d. with negligible variability, and the hypotheses are according to (3) and (4) we may adopt the permutation partial tests: $T_t^* = \sum_{j=1}^{C} [\overline{Y}_j^*(t)]^2 \cdot n_j$, $t = 1, ..., k$, where $\overline{Y}_j^*(t) = \sum_i Y_{ji}^*(t)/n_j$. These partial tests are permutationally equivalent to C-sample Snedecor's type statistics.

The same procedure applies also for the stochastic dominance problem (5). In this case a set of permutation partial tests might be: $T_t^* = \overline{Y}_2^*(t)$, $t = 1, ..., k$, which are permutationally equivalent to Student's type statistics:

$$[\overline{Y}_2^*(t) - \overline{Y}_1^*(t)] \, / \left\{ \sum_{ji} [Y_{ji}^*(t) - \overline{Y}_j^*(t)]^2 \right\}^{1/2}, \tag{6}$$

where most of constant coefficients are ignored. Sometimes, for compensating a possible weak heteroscedasticity between groups, Welch's type statistics may be used.

Observe that the stochastic dominance problems (5) admit very difficult parametric solutions when $k > 3$ (CHATTERJEE, 1984). On the contrary, within the permutation approach and through the nonparametric combination method the solution becomes easy and effective, although this procedure is rather computer intensive.

Extensions of the procedure to q-dimensional repeated observations ($q > 1$) and to missing values situations are possible under rather weak conditions, according to the approach developed in PALLINI & PESARIN (1995).

6 Some evaluations for the stochastic dominance test

To evaluate the behaviour of the nonparametric combination solution for the stochastic dominance testing problem, several Monte Carlo experiments have been carried out in a variety of situations under model (1) with baseline observations $Y_{ji}(0)$ and with stochastic additive effects. The dominance hypotheses under testing were according to (5). Some of these experiments considered the behaviour under H_0, some others the behaviour under H_1. The univariate data were generated under model (1) specified with: $C = 2$; $\sigma(t) = \sigma^{t-1}$; $Z_{ji}(t) \sim \mathcal{N}(0, 1)$ standard normal i.i.d; $Y_{ji}(0)$ i.i.d. $\mathcal{N}(\mu_j; 1)$. The autoregressive stochastic additive effects were generated according to an AR(1) submodel (2), with the following specifications: $n_1 = n_2 = 10$, $\mu_1 = \mu_2 = 100$, $\eta_j(t) = \eta_j^{t-1} - 1$, $\eta_1 = 1$, $\eta_2 = \eta \geq 1$, $\Delta_j(0) = 0$ and $\Delta_j(t) = \Delta_j(t-1) \cdot \nu^{t-1}$

$+ Q_3 \cdot \beta \cdot t$, and $Q_3 = \sum_{r=1}^{3} U_r$, i.e. the standardized sum of three i.i.d. random deviates $\mathcal{U}(-1,1)$ uniformly distributed in the $(-1,1)$ interval. Of course, when $\beta = 0$ the model is with fixed effects and when $\eta = 1$ the null hypothesis is true. In each simulation the number of resamplings was $B = 750$ and the number of Monte Carlo iterations was $MC = 2000$ under H_0 and $MC = 1000$ under H_1. Since in these experimental conditions the variability of $Y_{ji}(0)$ is not negligible, the baseline values were assumed with the role of covariates. Hence, the analyses were carried out on the increments with respect to baseline $X_{ji}(t) = Y_{ji}(t) - Y_{ji}(0)$. The test statistics we used was of the Student type (6) and the combination algorithm was the direct one.

The reported results show a very good agreement of estimated risks under H_0 with the nominal ones in all conditions. The behaviour under H_1 confirms the general expectations. In particular it is noticeable that the power increases when η or k increase and when σ or ν or β decrease.

	H_0				H_1							
σ	1.2	1.2	1.1	1.2	1.2	1.2	1.2	1.2	1.2	2	1.2	1.2
η	1	1	1	1	1.25	1.5	1.25	1.5	1.25	1.25	1.25	1.25
ν	1.1	1.1	1.2	2	1.1	1.1	1.1	4	1.1	1.1	4	1.1
β	.5	.005	.01	.1	.005	.005	.25	.005	.005	.05	.05	.1
k	3	3	6	6	3	3	3	3	6	6	6	6
α												
.010	.012	.009	.010	.012	.184	.787	.082	.285	.753	.039	.020	.395
.025	.026	.025	.023	.026	.293	.887	.158	.426	.872	.078	.051	.551
.050	.051	.049	.048	.048	.430	.949	.248	.577	.935	.137	.079	.687
.100	.102	.102	.097	.097	.591	.981	.382	.714	.970	.233	.149	.813
.200	.203	.197	.196	.199	.743	.991	.533	.865	.990	.379	.281	.902

Acknowledgments. This study have been carried out under a MURST financial support for researches on "Resampling methods of nonparametric inference".

References

1 Azzalini A. (1984). Estimation and Hypothesis Testing for Collection of Autoregressive Time Series. *Biometrika*, **23**, pp. 85-90.

2 Chatterjee S.K. (1984). Restricted Alternatives. In *Handbook of Statistics* **4**, P.R. Krishnaiah and P.K. Sen, Eds. Elsevier Science Publishers. North Holland, Amsterdam, pp. 327-345.

3 Crowder M.J., Hand D.J. (1990). Analysis of Repeated Measures. Chapman and Hall, London.

4 Davidian D., Giltinan D.M. (1995). Nonlinear Models for Repeated Measurements Data. Chapman and Hall.

5 Diggle P.J., Liang K.Y., Zeger S.L. (1994). Analysis of Longitudianl Data. Oxford Science Publications, Oxford.

6 FOLKS J.L. (1984). Combination of Independent Tests. In *Handbook of Statistics* **4**, P.R. Krishnaiah and P.K. Sen, Eds. Elsevier Science Publishers. North Holland, Amsterdam, pp. 113-121.

7 LAIRD N.M., WARE J.H. (1982). Random Effects Models for Longitudianl Data. *Biometrics,* **38**, pp. 963-974.

8 PALLINI A. (1994). Bahadur Exact Slopes for a Class of Combinations of Dependent Tests. *Metron,* **52**, No. 1-2, pp. 53-65.

9 PALLINI A., PESARIN F. (1995). Lecture Notes in Resampling Techniques for Testing Non Parametric Hypotheses. CLEUP, Padova.

10 PESARIN F. (1992). A Resampling Procedure for Non Parametric Combination of Several Dependent Tests. *Journal of the Italian Statistical Society,* **1**, pp. 87-101.

11 WARE J.H. (1985). Linear Models for the Analysis of Longitudianl Studies. *The American Statistician,* **39**, No. 2, pp. 95-101.

Adaptive Tests for Trend

MARKUS NEUHÄUSER
Solvay Pharmaceuticals
Hans-Böckler-Allee 20, D-30173 Hannover, Germany
LUDWIG A. HOTHORN
University of Hannover, LG Bioinformatik
Herrenhäuser Str. 2, D-30419 Hannover, Germany

Key Words: Trend test, unknown shape, adaptive concept

Abstract

In the nonparametric location model the Jonckheere (1954) trend test is powerful for convex shapes of the location parameters, whereas the Fligner and Wolfe (1982) test is powerful for concave shapes. Since in most situations knowledge about the shape is a priori lacking, we suggest an adaptive test which classifies the shape. We use the skewness of the pooled sample as a selector statistic. An adaptive trend test which classifies the shape is also possible in the parametric model where we consider contrast tests. In both models the adaptive trend test is – analogous to other adaptive tests – not optimal for any particular shape but relatively good over all shapes.

1 Introduction

Statistical methods which are routinely used in industrial applications should be robust concerning the underlying distribution and, in k-sample problems, the shape of the location parameters. For the sake of robustness concerning the distribution we firstly consider a nonparametric model, i.e. the location model in which there are one control and k treatment groups. Let $Y_{i1}, \ldots, Y_{in_i}$, $i=0, \ldots, k$, be $k+1$ independent random samples, with Y_{ij}, $j=1, \ldots, n_i$, having continuous distribution functions $F_i(t) = F(t - \theta_i)$, i.e. the distribution functions are the same except perhaps for a change in their location parameters. The index 0 belongs to the control group.

The null hypothesis is the equality of the location parameters, i.e. $H_0: \ \theta_0 = \theta_1 = \ldots = \theta_k$. We consider trend tests and, therefore, we restrict the alternative to the following one-sided ordered alternative hypothesis $H_A: \ \theta_0 \leq \theta_1 \leq \ldots \leq \theta_k$ with at least $\theta_0 < \theta_k$.

Under this alternative, different shapes are possible. In an extreme concave shape there are no differences between the treated groups, only the control group has a lower location parameter: $\theta_0 < \theta_1 = \theta_2 = \ldots = \theta_k$. In an extreme convex shape only the group with the highest dose differs from the other groups: $\theta_0 = \theta_1 = \ldots = \theta_{k-1} < \theta_k$. In most situations knowledge about the shape is lacking, i.e. the shape is a priori unknown.

There are different tests for the hypotheses mentioned above, we focus on the tests of Jonckheere (1954) and Fligner and Wolfe (1982), hereafter referred to J test and FW test, respectively. Other tests, e.g. the Cuzick (1985) test and the Le (1988) test, are not better than the considered tests (Mahrer and Magel, 1995).

Let U_{ij} be the Mann-Whitney statistic computed between the samples from the i-th and j-th populations. The J test statistic is $T_J = \sum_{i=0}^{k-1} \sum_{j=i+1}^{k} U_{ij}$, and the FW test statistic is $T_{FW} = \sum_{j=1}^{k} U_{0j}$. Hothorn (1989) compared the J and the FW test in terms of asymptotic relative efficiency. The J test shows a higher efficiency for convex shapes, and the FW test is more powerful for concave and umbrella shapes. According to simulation studies, these results also hold under finite situations.

Since the statistician usually has no information about the shape, an adaptive test would be useful. Hogg's adaptive concept (Hogg, 1974; Büning, 1991) is based on a decomposition of a so-called selector statistic which must be independent on all possible test statistics under H_0. The value of the selector statistic decides which test has to be conducted. The chosen test can be performed without α-adjustment. This adaptive concept is often used to classify the type of the underlying distribution with measures like tailweight or skewness and to select an appropriate rank test for the classified type of distribution. Recently, this approach was presented for the k-sample problem (Beier, 1996; Büning, 1996). We propose adaptive tests for trend which classify the shape both nonparametric (section 2) and parametric (section 3).

2 Adaptive Nonparametric Trend Test

In an adaptive trend test we want to classify the shape of the location parameters. In case that the underlying distribution is symmetrical, the distribution of the pooled sample is skewed to the right for convex shapes and skewed to the left for concave shapes. For example, let the control group be standard normal distributed and each treated group normal distributed with mean 1 and variance 1. Then the shape is concave and the coefficient of skewness of the pooled sample is $\frac{(k-k^2)/(k+1)^3}{(k/(k+1)^2+1)^{3/2}} < 0 \ \forall k \geq 2$. More detailed computations can be found in Neuhäuser (1996). Hence, we use the empirical skewness S of the pooled sample as a selector statistic. The adaptive test is defined as follows: If $S > 0$, conduct the J test, if $S \leq 0$, conduct the FW test.

Unfortunately, the selector S is useful only for symmetrical distributions. We therefore have to restrict the class of distribution functions accordingly.

We compared the J, the FW, and the adaptive test in a simulation study which has been performed with SAS version 6.10. For each combination of parameters, 10 000 simulation runs were generated. The following Table 1 contains representative results for various shapes and the common design with a control and three dose groups. The underlying distribution is a normal distribution with variance 1. For the sample size of 10 per group, the asymptotic normality of the J and the FW test statistic, respectively, is used for inference. For the smaller sample size of 5 per group, exact permutation tests have been performed. The adaptive test is of course not optimal for any particular shape. However, over all shapes, the adaptive test is relatively good since it does not lose as much power as a nonadaptive competitor for some shapes. It is just the philosophy of an adaptive test which works in the sense of the "safety first" principle. Consequently, the adaptive trend test is robust concerning the shape.

Table 1

Shape	Means	Power		
		J test	FW test	Adapt. test
Asymptotic tests (k=3, n_i=10, α=0.05, σ=1)				
no diff.	0, 0, 0, 0	0.05	0.05	0.05
linear	0, 0.5, 1, 1.5	0.95	0.81	0.89
convex	0, 0, 0, 0.5	0.26	0.11	0.19
	0, 0, 0, 1	0.62	0.19	0.44
	0, 0, 0, 1.5	0.87	0.28	0.70
	0, 0, 0, 2	0.97	0.35	0.88
	0, 0, 0, 2.5	0.99	0.40	0.96
concave	0, 1.5, 1.5, 1.5	0.87	0.99	0.95
umbrella	0, 1, 2, 1	0.79	0.96	0.87
Permutation tests (k=3, n_i=5, α=0.05, σ=1)				
no diff.	0, 0, 0, 0	0.05	0.05	0.05
linear	0, 0.5, 1, 1.5	0.75	0.53	0.65
convex	0, 0, 0, 1.5	0.58	0.15	0.45
concave	0, 1.5, 1.5, 1.5	0.58	0.84	0.75
umbrella	0, 1, 2, 1	0.47	0.73	0.60

The adaptive trend test can be used in drug development both in randomized clinical dose-finding trials and in preclinical studies. As an example the following Table 2 displays Ames assay data (Neuhäuser and Hothorn, 1997). An Ames assay is a mutagenicity test which is used as a preclinical test. The skewness is here S=1.16, therefore, the J test is chosen and leads to an exact one-sided P-value of 0.05. The P-value of the single FW test is larger (P=0.07).

Table 2

Dose (in μg)	Number of revertants				
0	16	17	17	20	18
5	18	18	19		
15	16	20	20		
50	20	24	28		
150	26	28	20		
500	16	20	16		

3 Adaptive Parametric Trend Test

Since we do not classify the type of the underlying distribution, we can construct an adaptive parametric trend test. We restrict the class of distributions to normal distributions, and we consider contrast tests. The Helmert contrast has the coefficients $-1, \ldots, -1, k$, the reverse-Helmert contrast the coefficients $-k, 1, \ldots, 1$. According to power computations and asymptotic efficiency results (Neuhäuser, 1996) the Helmert contrast test is powerful for convex shapes, and the reverse-Helmert contrast test for concave shapes. Therefore, we conduct the Helmert test if $S > 0$, and the reverse-Helmert test if $S \leq 0$. A simulation study showed that the power comparison is analogous to the nonparametric results mentioned above (Neuhäuser, 1996).

Further research on a comparison with multiple contrast tests and on an adaptive test which classifies both the shape and the underlying distribution is in progress.

References

Beier, F. (1996), *Adaptive Tests bei nicht-monotonen Dosis-Wirkungsbeziehungen*, University of Dortmund: Doctoral Thesis.

Büning, H. (1991), *Robuste und adaptive Tests*, Berlin: De Gruyter.

Büning, H. (1996), Adaptive tests for the *c*-sample location problem – the case of two-sided alternatives. *Commun. Statist. – Theory Meth.* A26, 1569-1582.

Cuzick, J. (1985), A Wilcoxon-type test for trend, *Stat. Med.* 4, 87-90.

Fligner, M. A.; D. A. Wolfe (1982), Distribution-free tests for comparing several treatments with a control, *Statistica Neerlandica* 36, 119-127.

Hogg, R. V. (1974), Adaptive robust procedures, *J. Amer. Stat. Assoc.* 69, 909-927.

Hothorn, L. A. (1989), On the behaviour of Fligner-Wolfe-trend test 'control versus *k* treatments' with application in toxicology, *Biom. J.* 31, 767-780.

Jonckheere, A. R. (1954), A distribution-free k-sample test against ordered alternatives, *Biometrika* 41, 133-145.

Le, C. T. (1988), A new rank test against ordered alternatives in k-sample problems, *Biom. J.* 30, 87-92.

Mahrer, J. M.; R. C. Magel (1995), A comparison of tests for the k-sample, non-decreasing alternative, *Stat. Med.* 14, 863-871.

Neuhäuser, M. (1996), *Trendtests bei a priori unbekanntem Erwartungswertprofil*, University of Dortmund: Doctoral Thesis.

Neuhäuser, M.; L. A. Hothorn (1997), The control of the consumer risk in the Ames assay, *Drug Inf. J.* 31, Issue 2.

Part VI

Computing, Imaging and Prediction

Convergence Rates of Simulated Annealing-Type of Recursive Algorithms

G. YIN*
Department of Mathematics, Wayne State University
Detroit, MI 48202

Abstract

This work is concerned with rates of convergence of simulated annealing type of algorithms for Monte Carlo optimization. We derive error bounds for the approximation sequences via the use of perturbed Liapunov function methods, and we show that a suitably scaled sequence converges weakly to a diffusion process.

Key Words: rate of convergence, annealing, Monte Carlo optimization.

1 Introduction

This paper is concerned with the algorithms

$$X_{n+1} = X_n - \frac{A}{n^\gamma}(\nabla f(X_n) + \xi_n) + \frac{B}{n^{\gamma/2}\sqrt{\ln[n^{1-\gamma} + A_0]}} W_n, \quad \text{for } 0 < \gamma < 1, \tag{1}$$

and

$$X_{n+1} = X_n - \frac{A}{n}(\nabla f(X_n) + \xi_n) + \frac{B}{\sqrt{n \ln\ln(n + A_0)}} W_n, \tag{2}$$

where A, A_0 and B are some positive constants, and the initial estimate X_1 is a bounded random variable (In fact, more general initial conditions can be treated; we assume the current condition for simplicity). These algorithms are a Monte-Carlo version of the "simulated annealing" procedures. Our work is motivated by the recent development of Kushner [9], Gelfand and Mitter [6], in which the convergence of the algorithms has been established. To the best of our knowledge, the convergence rates of these algorithms have not been studied up to date. This brings us to the current work.

*Research of this author was supported in part by the National Science Foundation under grants DMS-9529738 and DMS-9600245.

*Kitsos, C.P. and Edler, L., Eds.:Industrial Statistics, Physica-Verlag, Heidelberg, 1997

Many applications in control, optimization and related fields require one find the global minimizer of a given smooth function $f(\cdot) : \mathbb{R}^r \mapsto [0, \infty)$. Very likely, $f(\cdot)$ has multiple local minima. We are interested in the case that the gradient of $f(\cdot)$ cannot be readily calculated, only noise corrupted measurements $\nabla f(x)$+noise is available. One may try to use stochastic approximation type of algorithms to resolve the problem, but it is well known that a simple stochastic approximation algorithm may lead to the convergence to a local minimum (see for example, Kushner and Clark [10] or the more up to date and comprehensive treatment of Kushner and Yin [14]). Much effort thus has been devoted to the aforementioned problems. In the 80's, the so-called simulated annealing approach started attracting the attentions of researchers and practitioners. In [7], Kirkpatrick, Gelatt and Vecchi proposed a method of solution by running the Metropolis algorithm [16] while gradually lowering the temperature. Further analysis on the methods for recursive stochastic algorithms were contained in [6, 9] among others.

The study is closely related to randomly perturbed dynamic systems (see Freidlin and Wentzell [5]). Thus, studying the asymptotic properties of such algorithms often entails the investigation of the corresponding diffusion process. For instance, in [6], the convergence in probability of (2) was proved via the use of properties of diffusion processes (see Chiang, Hwang and Sheu [2]). Note that in algorithms (1) and (2), there are two noise sequences. One of them is the observation noise whereas the other is a sequence of added random disturbances. As a complement of the existing results, we focus on the convergence rate issues here, and show that the rates of convergence mainly depend on the perturbation noise not the observation noise. The rates are much slower than that of the stochastic approximation algorithms.

The rest of the paper is arranged as follows. The next section presents the assumptions needed together with the main results. Owing to the constraint of the page limitation, the verbatim proofs are omitted, although all conditions required are provided. Section 3 elaborates further on the results and mentions future research topics.

2 Main Results

To carry out the analysis, we make the following assumptions:

(A1) $f : \mathbb{R}^r \mapsto [0, \infty)$ is a three times continuously differentiable function such that

- $\min f(x) = 0$;
- There is a x^* that is the global minimum of $f(\cdot)$. Without loss of generality, assume $x^* = 0$ henceforth.

(A2) Suppose that for both algorithms (1) and (2), $X_n \xrightarrow{n} 0$ in probability.

(A3) Conditions on the noise sequences are:

(a) $\{W_n\}$ is a sequence of independent and identically distributed ($\mathbb{R}^r$-valued) random variables with $EW_n = 0$ and $EW_nW_n' = I$, where z' denotes the transpose of z for any $z \in \mathbb{R}^{r\times l}$ for some $l \geq 1$. $\{W_n\}$ is independent of $\{\xi_n\}$ and the initial estimate X_1.

(b) $E\xi_n = 0$ for each n, and $\sup_n E|\xi_n|^2 < \infty$.

(c) There exists a sequence $\{\rho(n)\}$ of nonnegative real numbers such that for $j \geq n$,

$$|E_n\xi_j - E\xi_j| \leq \rho(j-n) \text{ and}$$
$$\sum_{k=0}^{\infty} \rho(k) < \infty,$$

where E_n denotes the conditioning on $\mathcal{F}_n$, the σ-algebra generated by

$$\{X_1, \xi_j, W_j;\ j < n\}.$$

(A4) There is a twice continuously differentiable function $V(\cdot) : \mathbb{R}^r \mapsto [0,\infty)$ such that

(a) $V(x) \to \infty$ as $|x| \to \infty$.

(b) there is a $\lambda > 0$ such that $V_x'(x)\nabla f(x) > \lambda V(x)$ for all $x \neq x^*$, where $V_x(\cdot)$ denotes the first order partial derivatives (gradient) of $V(\cdot)$ with respect to x and similarly, $V_{xx}(\cdot)$ denotes the second order derivatives.

(c) $|V_x(x)| \leq K(1+V^{1/2}(x))$, $|V_{xx}(\cdot)|$ is bounded and $|\nabla f(x)|^2 \leq K(1+V(x))$.

Theorem 1. *Suppose that* (A1)–(A4) *are satisfied. Then the following assertions hold.*

(a) $EV(X_n) = O(1)$ *for* $\{X_n\}$ *defined by* (1) *and* (2).

(b) *For n large enough,*

$$\sup_n [\ln((n+1)^{1-\gamma} + A_0)]EV(X_{n+1}) = O(1), \quad \text{with } 0 < \gamma < 1$$

for Algorithm (1), *and*

$$\sup_n [\ln\ln((n+1) + A_0)]EV(X_{n+1}) = O(1)$$

for Algorithm (2), *respectively.*

Corollary 2. *If the Liapunov function $V(\cdot)$ is locally quadratic, i.e.,*

$$V(x) = x'Qx + o(|x|^2),$$

where Q is a symmetric positive definite matrix, then $\{v_n\}$ defined for both Algorithm (1) and Algorithm (2) is tight for sufficiently large n.

To proceed, define a sequence $\{v_n\}$

$$v_n = \begin{cases} \sqrt{\ln(n^{1-\gamma} + A_0)}X_n, & \text{for } 0 < \gamma < 1 \\ \sqrt{\ln\ln(n + A_0)}X_n, & \text{for } \gamma = 1. \end{cases} \tag{3}$$

Note that

$$\nabla f(x) = Hx + x' f_{xxx}(x^+)x,$$

where $H = f_{xx}(0)$ is the Hessian of $f(\cdot)$ at 0, and x^+ is a point on the line segment joining 0 and x. Linearizing (1) yields that

$$X_{n+1} = X_n - \frac{A}{n^\gamma}(HX_n + \xi_n) - \frac{A}{n^\gamma}X_n' f_{xxx}(X_n^+)X_n + \frac{B}{n^{\gamma/2}\sqrt{\ln(n^{1-\gamma} + A_0)}}W_n,$$

where X_n^+ is a point on the line segment joining 0 and X_n. Define

$$t_n = \sum_{i=1}^{n} \frac{A}{i^\gamma} \quad \text{and} \quad m(t) = \max\{n;\ t_n \le t\},$$

and the piecewise constant interpolation of v_n by

$$v^n(t) = v_{n+k} \text{ for } t \in [t_{n+k}, t_{n+k+1}),$$
$$v^n(0) = v_n.$$

Then $v^n(\cdot)$ is a sequence of random processes with paths in $D^r[0, \infty)$ the space of right continuous functions with left-hand limit endowed with the Skorokhod topology. A careful analysis leads to:

Theorem 3. *Suppose the conditions of Theorem 3.2 are satisfied. Then $\{v^n(\cdot)\}$ is tight in $D^r[0, \infty)$. Any weakly convergent subsequence of $\{v^n(\cdot)\}$ has a limit satisfying the following stochastic differential equation*

$$dv = -Hvdt + Cdw, \tag{4}$$

where $C = B/\sqrt{A}$, and $w(\cdot)$ is an r-dimensional standard Brownian motion.

3 Further Discussion

In view of the limit results obtained, the following points are worth noticing.

- The study of the global optimization algorithms appears to be closely related to the properties of the diffusion processes. It will be interesting to see the connection of such algorithms and singularly perturbed diffusions (see, for instance, Khasminskii and Yin [8], Yin and Zhang [18] and the references therein).

- Since $H = f_{xx}(0)$, $-H$ is a stable matrix. The stationary covariance Σ of the diffusion in (4) can be obtained by solving the Liapunov equation

$$\Sigma H + H'\Sigma = -C^2 I.$$

- A direct consequence of Theorem 3 and the observation above is that for both algorithms (1) and (2), $\sqrt{\ln(n^{1-\gamma} + A_0)}X_n$ and $\sqrt{\ln\ln(n + A_0)}X_n$ are asymptotically normal with zero mean and covariance Σ.

- The rates of convergence of the global stochastic approximation algorithms are mainly depending on the step size of the added random perturbations. Since the step size a_n is much smaller than b_n, the observation noise is averaged out in the limit.

- A consequence of the item above manifests that the step size in (1) is preferable. From a computation point of view, one always wants to choose large step size whenever possible to force the iterates approaching the desired target value faster. This is another reason that $a_n = O(1/n^\gamma)$ for $0 < \gamma < 1$ is more preferable compared with $a_n = O(1/n)$. On the other hand, one does not normally select a_n to be too large either since it will lead to more variation and oscillating behavior (see the discussion in Kushner and Yin [14, Chapter 10], and also [15] for related discussions).

- In the limit stochastic differential equation (4) for bother algorithms (1) and (2), the drift is exactly the same. This reveals another feature of the global stochastic optimization algorithms. For classical stochastic approximation algorithms, e.g., (1) and (2) without the added random perturbations, the scalings are $n^{\gamma/2}$ and $\sqrt{n}$, respectively. The corresponding drifts of the limit stochastic differential equations are $\widetilde{H} = H$ for $0 < \gamma < 1$ and $\widetilde{H} = H - I/2$ for $\gamma = 1$, respectively.

- One may wish to consider a related problem in which the observation $\nabla f(x)$+noise cannot be obtained, and only the observation $f(x)$+noise is available. In this case, the $\nabla f(x)$+noise in the algorithms must be replaced by a noisy finite difference estimate (such as in the Kiefer–Wolfowitz procedure).

- In various applications, we may encounter the situation that X_n are externally generated and cannot be chosen by us. In this case, it is useful to connect the stochastic approximation techniques with that of nonparametric methods (see Yin and Yin [17] for further references). Corresponding algorithms for the purpose of global optimization can be considered and the rates of convergence can be analyzed. It will be a combination of the methods provided in [17] and the current paper.

References

[1] A. Benveniste, M. Métivier and P. Priouret, *Adaptive Algorithms and Stochastic Approximation*, Springer-Verlag, Berlin, 1990.

[2] T.S. Chiang, C.R. Hwang and S.J. Sheu, Diffusion for global optimization in $\mathbb{R}^n$, *SIAM J. Control Optim.* **25** (1987), 737-752.

[3] S. Geman and C.R. Hwang, Diffusions for global optimization, *SIAM J. Control Optim.* **24** (1986), 1031-1043.

[4] S.N. Ethier and T.G. Kurtz, *Markov Processes, Characterization and Convergence*, Wiley, New York, 1986.

[5] M.I. Freidlin and A.D. Wentzell, *Random Perturbations of Dynamical Systems*, Springer-Verlag, Berlin, 1984.

[6] S.B. Gelfand and S.K. Mitter, Recursive stochastic algorithms for global optimization in $\mathbb{R}^d$, *SIAM J. Control Optim.* **29** (1991), 999-1018.

[7] S. Kirkpatrick, C.D. Gelatt and M.P. Vecchi, Optimization by simulated annealing, *Science* **220** (1983), 671-680.

[8] R.Z. Khasminskii and G. Yin, On transition densities of singularly perturbed diffusions with fast and slow components, *SIAM J. Appl. Math.* **56** (1996), 1794-1819.

[9] H. J. Kushner, Asymptotic global behavior for stochastic approximation and diffusions with slowly decreasing noise effects: global minimization via Monte Carlo, *SIAM J. Appl. Math.* **47** (1987), 169-185.

[10] H.J. Kushner and D.S. Clark, *Stochastic Approximation Methods for Constrained and Unconstrained Systems*, Springer-Verlag, 1978.

[11] H.J. Kushner, *Approximation and Weak Convergence Methods for Random Processes, with applications to Stochastic Systems Theory*, MIT Press, Cambridge, MA, 1984.

[12] H.J. Kushner and H. Huang, Rates of convergence for stochastic approximation type of algorithms, *SIAM J. Control Optim.* **17** (1979), 607-617.

[13] H.J. Kushner and G. Yin, Asymptotic properties of distributed and communicating stochastic approximation algorithms, *SIAM J. Control Optim.* **25** (1987), 1266-1290.

[14] H.J. Kushner and G. Yin, *Stochastic Approximation Algorithms and Applications*, Springer-Verlag, New York, 1997.

[15] P. L'Ecuyer and G. Yin, Budget-dependent rate of stochastic approximation, to appear in *SIAM J. Optim.*

[16] M. Metropolis, A. W. Rosenbluth, M.N. Rosenbluth, A.H. Teller and E. Teller, Equations of state calculations by fast computing machines, *J. Chem. Phys.* **21** (1953), 1087-1091.

[17] G. Yin and K. Yin, Passive stochastic approximation with constant step size and window width, *IEEE Transactions on Automatic Control*, **AC-41** (1996), 90-106.

[18] G. Yin and Q. Zhang, *Continuous–Time Markov Chains and Applications: A Singular Perturbation Approach*, Springer-Verlag, New York, 1997.

Spatial Statistics in the Material Research

Hynek Lauschmann, [1]Viktor Beneš

Department of Materials, Faculty of Nuclear Sciences and Physical Engineering, Czech Technical University, Trojanova 13, Karlovo nam. 13, 12000 Praha, Czech Republic
[1]Department of Mathematics, Faculty of Mechanical Engineering, Czech Technical University, 2, 12135 Praha 2, Czech Republic

Abstract:
Spatial statistics differs from classical statistics in many aspects, e.g. correlated data samples, edge effects, variety of sampling schemes. In the present paper we emphasize the two-stage character of spatial data evaluation. A standard approach is that one first transforms the observed image pattern and then does statistical inference and interprets the results. We present two classes of transformation both followed by regression. In the first application stereological unfolding is desired to reconstruct spatial data from planar sections. In the second case a grey-tone image from scanning electron microscopy is evaluated. Results of statistical description are desired in material research, namely in damage modelling of composite materials and fractography in aircraft research and nuclear power industry.

AMS 1991 subject classification: Primary 62H10, secondary 60G60

Keywords: Fractography, image analysis, regression, spatial statistics, stereology.

Introduction

Industrial statistics includes many topics. Among them spatial statistics appears in applications where planar and spatial data are to be evaluated.

*Kitsos, C.P. and Edler, L., Eds.:Industrial Statistics, Physica-Verlag, Heidelberg, 1997

Our contribution comes from material research, its aim is to describe damaging of materials leading to fracture. Typically the structure of material is observed by light and/or scanning electron microscopy and the image in a bounded window yields the input.

A metallographic section of a material specimen including reinforcement particles in a matrix is an example of binary image. Stereological methods (Stoyan et al.,1995) are used to unfold the true distribution of geometrical parameters of particles from the planar image measurement.

A perspective of a fracture surface is an example where a grey-tone image is observed. The transfer function between the altitude of spatial surface and the observed grey level is unknown but still the planar texture yields a valueable information. This is extracted by application of a morphological tranformation and subsequent statistical evaluation. These two examples demonstrate a general feature of a group of problems in spatial statistics. One observes an image in a bounded window which is first mathematically transformed and then statistical inference is done.

The first part of the present paper includes introduction to the classical stereological unfolding problem and a review of recent theoretical results in this field. A parametric model of particle damage initiation in a material structure is then studied and probability of particle cracking estimated. The second part includes an application from fractography (Nedbal et al., 1989; Lauschmann and Blahout, 1994). An original method based on bluring the image is suggested. It defines a parameter of the observed perspective of fracture surface which is directly related to the crack growth rate.

Stereological unfolding problem

A bounded closed convex set in Euclidean space R^d is called a particle. Let a fixed particle X be described by n real geometrical parameters $x_1, x_2, ..., x_n$. A sampling design is represented by a random hyperplane ρ with probability distribution Q on the parametric space of hyperplanes. Assume that the intersection $Y = X \cap \rho \neq \emptyset$, then Y is a random closed convex set in R^{d-1} called a particle section. Let $y_1, ..., y_m$ be geometrical parameters describing Y, such that $y_1, ..., y_k$, $k \leq \min(n, m)$ correspond to properties of $x_1, ..., x_k$, e.g. x_1, y_1 size, x_2, y_2 shape factor etc. Let $p(y_1, ..., y_m | x_1, ..., x_n, \uparrow)$ be the conditional probability density of $y_1, ..., y_m$ given $x_1, ..., x_n$ and given that the particle is hit by ρ. The upper arrow $\uparrow$ emphasizes that the distribution p depends not only on particle characteristics but also on the sampling design Q.

Assume that particles are randomly dispersed in R^d with constant intensity N_d. For fixed particles (just translates of X) denote $N_{d-1}(x_1, ..., x_n)$ the mean (with respect to Q) intensity of particle sections in ρ. Further given N_d let particles be random with probability density $f(x_1, ..., x_n)$ of parameters $x_1, ..., x_n$ invariant with respect to translations in R^d. We are interested in particle sections observed in ρ. The average particle section intensity N_{d-1} and probability density $g(y_1, ..., y_m)$ of parameters $y_1, ..., y_m$ are defined by

$$N_{d-1}g(y_1, ..., y_m) = \tag{1}$$

$$\int .. \int N_{d-1}(x_1, ..., x_n)p(y_1, ..., y_m|x_1, ..., x_n, \uparrow)f(x_1, ..., x_n)dx_1..dx_n.$$

The stereological unfolding problem consists in the estimation of unknown particle characteristics f, N_d from the particle section distribution g and N_{d-1}, which can be observed and estimated from realizations of ρ. The first part of the solution is to establish the theoretical relations.

Proposition 1 *The unfolding problem is decribed by an equation*

$$N_{d-1}g(y_1, ..., y_m) = N_d \int .. \int k(x_1, ..., x_n, y_1, ..., y_m)f(x_1, ..., x_n)dx_1..dx_n, \tag{2}$$

for some nonnegative kernel function k.

Putting $k(x_1, .., x_n, y_1, ..., y_m) = \frac{N_{d-1}(x_1,...,x_n)}{N_d}p(y_1, ..., y_m|x_1, ..., x_n, \uparrow)$ we get this formula from definition.

The problems (2) were studied until now for m, n at most 2. They lead to Abel type integral equations which can be solved analytically, cf. Stoyan et al.(1995) for a review of spherical and ellipsoidal particles. For situations when the input data form a bivariate histogram, equation (2) is discretized and solved by numerical methods. Ohser and Mücklich(1995) use EM-algorithm to unfold the size-shape, size-number (number of vertices) distribution of various particular polyhedra. They demonstrate that even in some cases with complex section classification, when the form of (2) is unknown, the coefficients of its discrete version can be obtained by simulations. On the same theoretical basis a software package Stereologie under IMAGE C system products of IMTRONIC Gmbh was developed. It enables solution of many important unfolding problems in metallography, but all excluding orientations of particles.

The orientation of elongated particles is an important parameter, in earlier papers isotropy (uniform orientation distribution) was assumed. The bivariate size orientation problem including colatitudes was solved by Gokhale(1996) for circular plates and vertical uniform random (VUR) sections. This popular sampling design consists of random sections parallel to a given axis. Finally in Beneš et al.(1997) the unfolding of the trivariate size-shape-orientation problem for either oblate or prolate spheroids was derived. The idea of theoretical solution is as follows: in the vertical uniform random sampling design given a fixed particle the size distribution of its section is conditionally independent of shape and orientation, which enables to express $p(y_1, ..., y_m|x_1, ..., x_n, \uparrow)$ in product for and decompose the integral equation (2) in a series of two simpler ones. On the other hand in the remaining shape-orientation problem both parameters are conditionally functionally dependent and thus Frechet bounds from probability theory enter in the solution. Numerical evaluation of the desired trivariate distribution is based on discretization and the EM-algorithm. The stability of solution is described in terms of condition numbers of matrix operator corresponding to the discrete solution.

Modeling and estimating the particle damage

Further we will discuss data evaluation after the stereological procedure in a practical example concerning damage modeling of materials. Damage modeling is an important tool for the development of new high strength dispersion hardened alloys and metal matrix composites. Consider a model material of AlSi composite and uniaxial tensile tests carried out. Brittle silicon particles embedded in the ductile aluminium matrix do not deform plastically and particle cracking has been expected to occur during deformation. Previous experimental studies indicated that a relationship exists between incidence of cracking and particle morphological parameters (Brechet et al., 1991). This relationship was modeled by an application of Weibull statistics like in many papers on this topic. When carrying out this analysis authors assume that size and shape of particles are independent, and use a simplified product form of their joint distribution.

Our approach was first to use the stereological procedure to estimate the trivariate size-shape-orientation distribution. Flat particles are modeled by oblate spheroids, size is the length of main axis and shape the ratio between smaller and main axis, orientation the colatitude to vertical axis. The VUR sampling design is natural when taking probes parallel to the tensile test axis. The dependence between size and shape is observed in the material. The probability of fracture P of a particle of volume V, orientation θ and shape factor s is modeled by the Weibull statistics $P = 1 - \exp(-V/V_0[\sigma_p(s, \theta, \varepsilon, V)/\sigma_0]^m)$, where σ_0, V_0 are constants, m the Weibull modulus, ε the applied strain and σ_p the stress in the particle. A stress function $\sigma_p(s, \theta, \varepsilon, V) = \sigma_m(1 + a/6s)\sin\theta$ is recommended, where σ_m is the matrix tensile strength.

The idea is to estimate the probability P by means of relative frequency of cracked particles observed in the samples. Moreover, doing this in each ijk–th subclass of particle size-shape-orientation histogram, we get data for the estimation of m using non-linear regression techniques. The material scientists prefer the Weibull statistics from physical assumptions, which need not be fulfilled in complex materials. From the analogy with survival analysis a statistician may expect better fit when using the logistic regression. The evolution of the damage process was monitored measuring specimen at several levels of strain. As expected the parameter m did not significantly change (estimated values between 1 and 3), time series of further geometrical parameters is presented in Fig. 1.

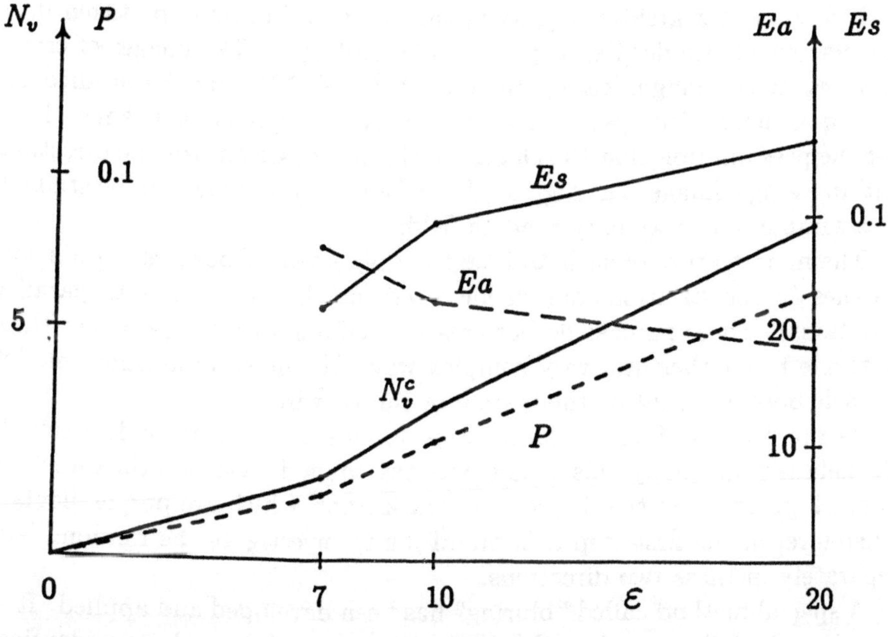

Fig.1: The evolution of particle cracking in the AlSi material. Ea [μm], Es is the mean size and mean shape factor of cracked particles, N_V^c [$10^6\ \mu m^{-3}$] their intensity and $P = \frac{N_V^c}{N_V}$ the global probability of particle fracture.

Quantitative fractography

Another important problem in material research is the fractographic reconstruction of the history of fatigue cracks in metallic construction parts. That means first to assess the crack growth rate in representative localities

on the fracture surface; secondly to construct the shape of the fracture as a function of time by means of integrating the crack velocity along crack growth direction.
We describe a new statistical method of solution, using the automatic image analysis.

Quantitative fractography provided a reliable approach based on analysis of striations. Striations are the finest traces of the fracture process in the crack surface. To investigate them magnification from about 5.10^3 to 5.10^4 must be used. When the crack had been affected by corrosion, errosion, etc., the fine striation relief was taken off. The task is concerned with identifying another fractographic feature which is closely correlated with the crack growth rate.

It follows that greater objects in the crack surface must be taken into account which are deep enough not to be destroyed. The images of crack surfaces in the magnification between 10^2 and 2.10^3 can be studied as a random field with specific texture. Magnifications in this range allow yet the presumption that the change of the crack rate and/or the direction within a single image is negligible. Therefore, a single image can be studied as a sample of a stationary random field.

The main feature of such an image is the system of borders of striation patches (if the striations are visible, every patch is an area with parallel striations). These patches do not create a cellular structure, because they pass one to another in a very complex way. The most appropriate model of their borders could be the system of curved wires.

As the process of the origin of crack surface is significantly directional, the field is principially anisotropic with two main directions - the direction of crack growth and the direction of crack front, which are perpendicular. Therefore, in the first step it is promising to investigate the random field separately in these two directions.

A special method called "bluring" has been developed and applied. It is based on the following idea: When averaging two, three and more identical images, each of which is shifted from the foregone by a constant in the same direction, the features of the process will gradually disappear and the brightness in each point will tend to the mean grey level. The measure of bluring effect is the contrast of the resulting image expressed by the grey level variance D. The function $D(n)$, i.e. the variance of blured image as a function of number of averages, is an important parameter of the image. In a simple monotonically decreasing form it summarizes the character of all special objects from the whole image, regarding the distances in the direction of shifting. A general regression form of the dependence $D(n)$ has been found with a dominant parameter γ_x.

The image bluring method

The theoretical background of the suggested method is the following.

Proposition 2 *Let X_t, $t = 1, 2, \ldots$ be a stationary process with mean zero and $var X_t = 1$ such that the autocorrelation function $\rho_X(h) = EX_tX_{t+h}$ satisfies $\rho_X(h) = 0$, $h > d$ for some $d > 0$. Put $Y_j^{(n)} = \frac{1}{n+1}\sum_{k=0}^{n} X_{j+k}$, $j = 1, 2, \ldots$ and denote $D(n) = var Y_j^{(n)}$. Then there exists a constant $\gamma_x > 0$ such that $\lim_{n\to\infty}(n+1)D(n) = \frac{1}{\gamma_x}$.*

Proof: It holds $EY_j^{(n)} = 0$,

$$D(n) = \frac{1}{(n+1)^2}\sum_{k=0}^{n}\sum_{l=0}^{n} EX_{j+k}X_{j+l} =$$

$$= \frac{1}{(n+1)^2}\sum_{k=0}^{n}\sum_{l=0}^{n}\rho_x(k-l) = \frac{1}{(n+1)^2}(n+1+2\sum_{p=1}^{n}(n+1-p)\rho_x(p)).$$

Thus

$$(n+1)D(n) = 1 + 2\sum_{p=1}^{n}\frac{n+1-p}{n+1}\rho_x(p) \to 1 + 2\sum_{p=1}^{d}\rho_x(p) = \frac{1}{\gamma_x}.$$

It is known that $\rho_x(h) = 0$, $h > d$ for some $d > 0$ if and only if X_t is a moving average process. Even if the assumptions of the proposition are not fulfilled, an analogue of parameter γ_x can be defined as the slope of asymptote of a regression curve for reciprocal sampling variances $\frac{1}{D(n)}$.

In the fractographic application realization of the process X_t is formed by grey level values along the crack front. This process is only locally stationary, typically we can work with local autocorrelation function of a single argument on intervals much larger than the correlation length $d = max_h|\rho_x(h)| > \varepsilon$ for a small $\varepsilon > 0$. A series of estimators of parameter γ_x is then obtained in the direction of crack growth and related to the physical process.

An application has been carried out on fracture surfaces of fatigue cracks of samples from the steel EDF6 used in nuclear power industry (see Fig.2). Bluring in the crack front direction (along x-axis) resulted into an exponential relation between parameter γ_x and the crack growth rate (see Fig.3). The variance was comparable with the corresponding relation from analysis based on striations. This result makes possible the crack history reconstruction also in the case when the striation relief has been damaged by an additional process.

The method using bluring can be generally used when investigating hardly detectable differences between random fields with a complicated structure. Complementary information about correlation properties of the image can be extracted from the regression curve besides the asymptote. Further research will be devoted to the case when there are substantial changes in the local autocorrelation functions along the crack front.

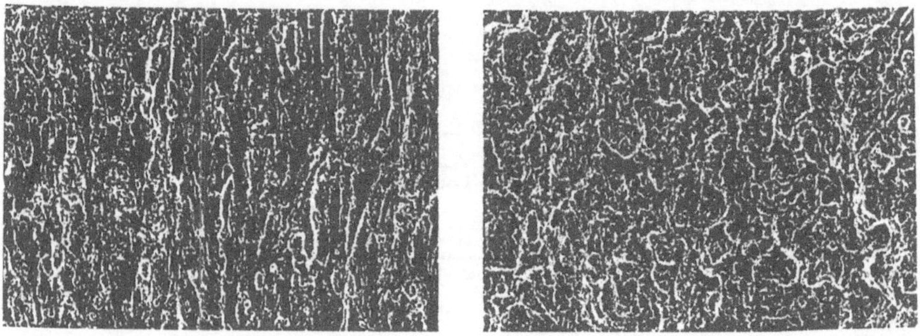

Fig.2: Images of crack surface corresponding to small (left) and high (right) crack growth rate.

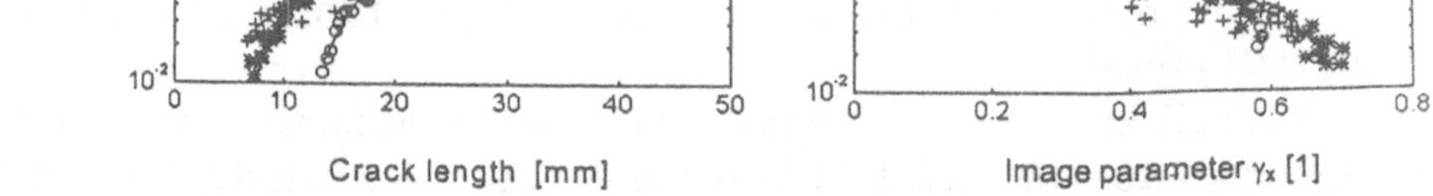

Fig.3: Three cracks with different course of crack growth rate (left). The correlation between the crack growth rate and the parameter γ of crack surface images (right).

Acknowledgement

The financial support of Grant Agency of the Czech Republic, projects no. 106/95/1482 and 106/97/0827, is greatly appreciated. We thank to Margarita Slámová, Research Institute for Metals, Panenské Břežany for the cooperation in the first problem on damage initiation.

References

Benes V., Jiruse M., Slamova M.(1997) Stereological unfolding of the trivariate size-shape-orientation distribution of spheroidal particles with application. Acta Mater., in press.

Brechet Y., Embury J.D., Tao S., Luo L.(1991) Damage initiation in metal matrix composites. Acta Metall. Mater., 39/8, 1781-1786.

Gokhale A.M.(1996) Estimation of bivariate size and orientation distribution of microcracks. Acta Mater., 44/2, 475-485.

Lauschmann H., Blahout M.(1994) A fractographic reconstitution of a fatigue crack front. Fatigue Fract. Engng Mater. Struct., 17/12, 1391-1396.

Nedbal I., Siegl J., Kunz J.(1989) Relation between striation spacing and fatigue crack growth rate in Al-alloy sheets. In: Advances in Fracture Research, Proc. ICF 7, Houston, Pergamon Press, Oxford, Vol.5, 3843-3849.

Ohser J., Mücklich F.(1995) Stereology for some classes of polyhedrons. Adv. of Appl. Probab., 27/2, 384-397.

Stoyan D., Kendall W.S., Mecke J.(1995) Stochastic Geometry and its Applications. 2nd Edit., Wiley, New York.

Exact Solutions to the Problem of Predicting a vast Class of Weakly Stationary, Linearly Singular, Discrete-Parameter Stochastic Processes

GIORGIO CELANT

Dipartimento di Scienze Statistiche – Università di Padova

Via S.Francesco, 33 – 35121 Padova – Italy

1 Introduction

It is well-known that a weakly stationary discrete-parameter stochastic process, X_t with $t \in \mathbf{Z}$ (set of integers), is the sum of two components: one is regular (R_t) and the other is singular (S_t) (Wald's theorem). In statistical literature, the following hypothesis is given: $S_t = 0, \forall\, t$. This hypothesis is arbitrary because it is easy to show the existence of singular processes and furthermore we do not have at our disposal a statistical test in order to state, looking at the data, if $X_t = R_t$ or $X_t = R_t + S_t$ or $X_t = S_t$. In order to prove that a singular process exists, we need to construct it. The following processes are singular:

- $X_{t+1} = \sum_{n=0}^{\infty} \sum_{k=0}^{n} \binom{n}{k} (-1)^k X_{t-k}$;
- $X_t = \sum_{i=1}^{\infty} (\xi_i \cos \lambda_i t + \eta_i \sin \lambda_i t)$, $\lambda_i \in (-\pi, \pi]$;
  $t \in \mathbf{Z}$, $E(\xi_i) = E(\eta_i) = 0$, $E(\xi_i^2) = E(\eta_i^2) = \sigma^2_{\lambda_i}$, $E(\xi_i \eta_j) = E(\xi_i \xi_j) = E(\eta_i \eta_j) = 0$, $\forall i \neq j$.

Hence, the problem is to examine singular processes in order to point out the properties useful to solve the classical inferential problems, as for example, the construction of a test to verify the presence of the singular component. It is important to verify the presence of the singular component because only if

*Kitsos, C.P. and Edler, L., Eds.:Industrial Statistics, Physica-Verlag, Heidelberg, 1997

$S_t = 0$, then the Box-Jenkins approach to time series analysis is statistically rigorous.

In this paper we propose a first step in the characterization of singular processes, i.e. we determine some linear representations. We describe the exact formula for a predictor related to a weakly stationary, linearly singular, discrete–parameter stochastsic process, whose spectrum $F(\lambda)$ (of derivative $F'(\lambda)$) satisfies the condition that $F'(\lambda) = 0$ within a subset of $(-\pi, \pi]$ with a non–zero Lebesgue measure.

2 Singular Processes: Definitions and characteristics

From a formal point of view, a weakly stationary, discrete–parameter stochastic process is a sequence of random variables, defined in a common probability space (Ω, B, P) and satisfying the two conditions:
1) $E(X_t) = \mu, \quad \forall t \in \mathbf{Z}$ (μ is later assumed equal to zero); and
2) $B(s,t) = B(s-t) = \int_{-\pi}^{\pi} e^{i\lambda(s-t)} dF(\lambda), \quad \lambda \in (-\pi, \pi]$.

$F(\lambda), \lambda \in (-\pi, \pi]$ denotes the inverse Fourier transform of $B(s-t)$. $F(\lambda)$ is defined as the spectral function of a process or, more simply, its spectrum. An interesting category of processes is that consisting of linearly singular processes.

The existence of singular processes is a consequence of Wold's decomposition theorem. Practical examples of these can be found in Celant (1983). An important property of weakly stationary, singular, discrete–parameter stochastic processes is that the Radon–Nikodỳm derivative of their spectrum ($F'(\lambda)$), calculated with respect to the Lebesgue measure, satisfies the equation: $\int_{-\pi}^{\pi} \ln F'(\lambda) d\lambda = -\infty$. This equation is known as the Kolmogorov theorem and provides a necessary and sufficient condition for the linear singularity of a process.

3 Analytical properties of the process

Stone's theorem (also known as the spectrum decomposition theorem for unit operators (Schaefer, 1980)) when applied to unit operator U^t, obtained by means of a continuous linear extension of the map defined by $\hat{U}^t : (X_t) \to (X_t)$, $X_k \mapsto X_{t+k}$, states that the process generated by $U^t(X_0) = X_t$ is represented by the following integral:

$$X_t = \int_{-\pi}^{\pi} e^{i\lambda t} dW(\lambda) \tag{1}$$

where $W(\lambda)$ is a continuous, quadratic-mean, independent increment stochastic process such that $W(-\pi) = 0$.

Equation (1) establishes the association between (X_t) and function $e^{i\lambda t}$. Let us therefore assume that: $\Phi(X_t) = e^{i\lambda t}$. In general, to an element of $H(X)$, that is, a function of the type: $g(X) = \sum_{t=-\infty}^{+\infty} a_t X_t$, we associate the function

$$\Phi(g(X)) = \sum_{t=-\infty}^{+\infty} a_t \Phi(X_t) = \sum_{t=-\infty}^{+\infty} a_t e^{i\lambda t}\,.$$

It can be demonstrated that function Φ defines an isometric isomorphism between $H(X)$ and $L^2_{\mathbf{Z}}(dF)$, where $H(x)$ is the Hilbert space generated by X_t and $L^2_{\mathbf{Z}}(dF)$ represents the Hilbert space generated by $\{e^{i\lambda t}\}$, $t \in \mathbf{Z}, \lambda \in (-\pi, \pi]$, that are summable–square functions with respect to $F(\lambda)$. This corrispondence enables interpretation of any problem relating to $H(X)$ in $L^2_{\mathbf{Z}}(dF)$. For example, to predict component X_t of process (X_t) with the set X^- (where $X^- = \{X_{t-1}, X_{t-2}, \ldots\}$), it is sufficient to express function $e^{i\lambda}$ as a "linear combination" $g(\lambda)$ of functions $\{e^{i\lambda n}\}$, $n \leq 0$. Owing to the fact that $\Phi^{-1}(e^{i\lambda}) = X(1)$ and $U^t(X(1)) = X_{t+1}$, this means that:

$$X_{t+1} = U^t\left(\Phi^{-1}\left(\sum_{j=0}^{+\infty} a_j e^{-i\lambda j}\right)\right)\,.$$

Therefore, assuming that $z = e^{i\lambda}$, elements a_j are the same as those of the series: $\frac{1}{z} = \sum_{j=0}^{+\infty} a_j z^j$. This series converges in $L^2_{\mathbf{Z}}(dF)$ and is therefore to be intended as converging with respect to measure dF.

In the demonstration that follows, it is assumed that the process is such that: $\exists F'(\lambda)$, $\exists K' \subsetneq (-\pi, \pi]$ such that $F'(\lambda) = 0$, $\forall \lambda \in K'$ and that K' has a non–zero Lebesgue measure.

Every set K' of this kind is necessarily associated to a linearly singular process, since $\int_{-\pi}^{\pi} \ln F'(\lambda) d\lambda = -\infty$. The existence of K' implies that (following Runge's theorem) there are a number of elements a_j (depending on K') which make the series $\sum_{j=0}^{+\infty} a_j z^j$ converge uniformly to $\frac{1}{z}$ in $(-\pi, \pi]\backslash K'$ and therefore, as uniform convergence implies converge in measure, this means that the series converges over the entire interval $(-\pi, \pi]$. In fact, uniform convergence of the series in $(-\pi, \pi] \setminus K'$ ensures convergence in measure in $(-\pi, \pi] \setminus K'$. Instead, convergence in measure in K' is a conseguence of the fact in K', $F'(\lambda) = 0$, $\forall \lambda \in K'$. The Lebesgue measure of K' is identified with the symbol $\varepsilon \in (0, 2\pi)$.

In the following section, elements a_j for a couple of simple situations are calculated, before passing on to a more general result in the final section.

4 Calculation of coefficients for a few simple cases

Two cases, a) and b) are distinguished

a) By definition, we know that $\frac{z}{z_0} = \sum_{n=0}^{+\infty}(-1)\frac{(z-z_0)^n}{z_0^n}$ converges uniformly for all $z \in I^0$, where I^0 is a subset of $I = \{z \in \mathbf{C} \setminus \{0\} : |z - z_0| \leq z_0\}$, and **C** is the set of complex numbers.
Remaining within the domain of complex numbers of unit norm, it may be stated that the series $e^{i\lambda} = \sum_{n=0}^{+\infty}\sum_{k=0}^{n}\binom{n}{k}(-1)^k e^{i\lambda k}$ converges uniformly in $I^0 = [\frac{-\pi}{3} + \lambda_0, \frac{\pi}{3} - \lambda_0]$, and therefore converges in measure in $(-\pi, \pi]$, assuming that $K = (-\pi, \pi] \setminus I^0$.
The process associated to this series is expressed by the following formula:

$$X_{t+1} = \sum_{n=0}^{\infty}\sum_{k=0}^{n}\binom{n}{k}(-1)^k X(t-k)\,.$$

b) Assuming that $\lambda \in [-\lambda_0, \lambda_0]$, $\lambda_0 \in (0, \frac{\pi}{2})$, then $\frac{1}{z}$ may be approximated as follows:
due to the fact that $\frac{1}{z} = \int_0^{\infty} e^{-tz}dt$ is an integral that converges rapidly for $\Re(z) \geq \varepsilon > 0$ ($\Re$ =Real component), then

$$\frac{1}{z} = \sum_{n=0}^{\infty}\frac{(-1)^n\Gamma^{n+1}}{(n+1)!}z^n + Er(\Gamma)\,.$$

Here, error $Er(\Gamma) = \frac{1}{z}e^{-\Gamma_z}$ tends rapidly to zero as $\Gamma \to +\infty$. Therefore, for $\lambda \in [-\lambda_0, \lambda_0]$, $\lambda_0 \in (0, \frac{\pi}{2})$ the following formula is obtained:

$$e^{i\lambda} \approx \sum_{n=0}^{\infty}\frac{(-1)^n\Gamma^{n+1}}{(n+1)!}e^{-i\lambda n}\,, \qquad \Gamma \in (0, +\infty)\,.$$

That is:

$$X(t+1) \approx \sum_{n=0}^{\infty}\frac{(-1)^n}{(n+1)!}\Gamma^{n+1}X(t-n)\,,$$

where Γ is determined by the condition that $|\frac{1}{z}e^{-\Gamma_z}| < \varepsilon$, for a given value of ε. In this case, $K' = (-\pi, \pi] \setminus (\frac{-\pi}{2}, \frac{\pi}{2})$, that is, $F'(\lambda) \neq 0$ for $\lambda \in (\frac{-\pi}{2}, \frac{\pi}{2})$ or any subset of the interval $(\frac{-\pi}{2}, \frac{\pi}{2})$.

5 Calculation of coefficients in a general case

Let $K' = (-\pi.\pi] \setminus K = \{z \in \mathbf{C} : |z| = 1, \quad -\theta_0 \leq \arg z \leq \theta_0,\ \theta_0 \in (0, \pi)\}$. We expand $\frac{1}{z}$ in K substituting $\frac{1}{z}$ with a rational function (introducing an error that may be made as small as we wish) having a pole outside the disc $|z| \leq 1$, and then consider its Taylor expansion centred on the origin, introducing another error (that we may make as small as we wish).

Let us take $a > 1$ and consider the real line interval $[-a, 0]$. The distance between this line and K is given by $2r = \sin(\pi - \theta_0)$, (for $\theta_0 > \frac{\pi}{2}$), and 1 if $\theta_0 \leq \frac{\pi}{2}$. Let us assume that $a = N \cdot r$. A sequence of points defined by $\alpha_n = nr$, $n = 0, 1, \ldots, N$ is introduced. The approximations of $\frac{1}{z}$ in K may be obtained by recursion, as follows; firstly, we attempt to evaluate $f_1(z)$, expressed by the formula

$$f_1(z) = \frac{1}{z}\left[1 - \left(\frac{\alpha_1}{\alpha_1 - z}\right)^{n_1}\right] = \left[\frac{(\alpha_1 - z)^{n_1} - \alpha_1^{n_1}}{z(\alpha_1 - z)^{n_1}}\right].$$

Distance $|\alpha_1 - z|$, $z \in K$ has a lower bound of $2r$, hence $|\frac{\alpha_1}{\alpha_1 - z}| \leq \frac{r}{2r} = \frac{1}{2}$.

Considering that $|\frac{1}{z}| = 1$ in K, this means that $|\frac{1}{z} - f_1(z)| < |\frac{\alpha_1}{\alpha_1 - z}| < 2^{-n_1}$ in K. Moreover, function $f_1(z)$ has a single pole in α_1 of n_1–th order, due to the fact that

$$(\alpha_1 - z)^{n_1} - \alpha_1^{n_1} = \sum_1^{n_1} \binom{n_1}{h} (-1)^h z^h (\alpha_1)^{n_1 - h} = z \sum_1^{n_1} \binom{n_1}{h} r^{n_1 - h} (-1)^h z^{h-1}.$$

Therefore,

$$f_1(z) = \frac{(-1)^{n_1} \sum_1^{n_1} \binom{n_1}{h} r^{n_1-h} (-1)^h z^{h-1}}{(z - \alpha_1)^{n_1}} = \frac{P_1(z)}{(z - \alpha_1)^{n_1}}.$$

Here, it is assumed that $P_1(z) = (-1)^{n_1} \sum_1^{n_1} \binom{n_1}{h} r^{n_1-h} (-1)^h z^{h-1}$. The second step involves introducing the following rational function:

$$f_2(z) = \frac{P_1(z)}{(z - \alpha_1)^{n_1}} \cdot \left[1 - \left(\frac{\alpha_1 - \alpha_2}{z - \alpha_2}\right)^{n_2}\right]^{n_1}.$$

As $(z - \alpha_2)^{n_2} - (\alpha_1 - \alpha_2)^{n_2}$ features a simple zero for $z = \alpha_1$, and as

$$|f_2 - f_1| \leq \max_K |f_1| \left\{1 - \left(1 - \left(\frac{1}{2}\right)^{n_2}\right)^{n_1}\right\} \leq 2^{n_1 - n_2} + 2^{-n_2},$$

the error committed is given by the following equation:

$$\left|\frac{1}{z} - f_2\right| \leq \left|\frac{1}{z} - f_1 + f_1 - f_2\right| \leq \left|\frac{1}{z} - f_1\right| + |f_2 - f_1| \leq 2^{-n_1} + 2^{n_1 - n_2} + 2^{-n_2}.$$

Proceeding recursively in the same manner, that is, by substituting function $f_h(z) = \frac{P_h(z)}{(z-\alpha_h)^{n_h}}$ with the rational function $f_{h+1}(z) = f_h(z) \cdot \left[1 - \left(\frac{\alpha_h - \alpha_{h+1}}{z - \alpha_{h+1}}\right)^{n_{h+.}}\right]^{n_h}$, after N substitutions the following approximation is obtained:

$$f_N(z) = \frac{(-1)^{n_1}\sum_{k=1}^{n_1}\binom{n_1}{k} z^{k-1}(-1)^k\left[\frac{\sin(\pi-\theta_0)}{2}\right]^{n_1-k}}{(z-\alpha_1)^{n_1}}.$$

$$\cdot\prod_{i=1}^{N-1}\left(1-\left(\frac{\alpha_i-\alpha_{i+1}}{z-\alpha_{i+1}}\right)^{n_{i+1}}\right)^{n_i}.$$

The error introduced by this approximation in K is given by:

$$\left|f_N-\frac{1}{z}\right| \le \sum_{i=2}^{N}|f_i-f_{i-1}|+\left|\frac{1}{z}-f_1\right| \le \sum_{i=2}^{N}(2^{n_{i-1}-n_i}+2^{n_i})+2^{-n_1}.$$

At this point, in order to obtain the desired polynomial approximation, it is necessary to find the Taylor expansion of $f_N(z)$ centred on the origin. As the singularity of f_N is $-a$, which represents a point situated outside the disc of unit radius $|z|\le 1$, it may be stated that the Taylor expansion of this function converges uniformly in K.

Let us now look at the coefficients of this expansion. By denoting the n–th derivative of $f_N(z)$ at $z=0$ by $D^n_{z=0}$, we have:

$$D^n_{z=0}(f_N(z)) = \sum_{h_1=0}^{n}\sum_{h_2=0}^{n-h_1}\cdots\sum_{h_{N-1}=0}^{n-(h_1+\cdots+h_{N-2})}\binom{n}{h_1}\binom{n-h_1}{h_2}\cdots\binom{n-\sum_{i=1}^{N-2}h_i}{h_{N-1}}$$

$$\cdot\prod_{j=1}^{N-2} D^{h_{j+1}}_{z=0}(g_j(z))D^{n-\sum_{i=1}^{N-1}h_i}_{z=0}(g_{N-1}(z))D^{h_1}_{z=0}(f_1(z))$$

where $g_i = \left(1-\left(\frac{\alpha_i-\alpha_{i+1}}{z-\alpha_{i+1}}\right)^{n_{i+1}}\right)^{n_i}$ $i=1,\ldots,N-1$ and $\alpha_i-\alpha_{i+1}=-r$. By direct calculation, it can be demonstrated that:

$$D^k_{z=0}(g_i) = \begin{cases} \sum_{j=0}^{n_1}\binom{n_i}{j}(-1)^{k-jn_{i+1}}(\alpha_i-\alpha_{i+1})^{jn_{i+1}}(\alpha_{i+1})^{-jn_{i+1}-k}. \\ \quad\cdot\prod_{s=0}^{k}(s+jn_{i+1}) & k>0 \\ \left(1-\left(\frac{\alpha_i-\alpha_{i+1}}{-\alpha_{i+1}}\right)^{n_{i+1}}\right)^{n_i} & k=0 \end{cases}$$

$$D^k_{z=0}(f_1(z)) = \begin{cases} \frac{-n_1}{\alpha_1} & k=0 \\ (-1)^{k+1}\sum_{h=0}^{n_1-1}\frac{1}{\alpha_1^{k+1}}\prod_{j=0}^{k}(h-n_1-j) & k\neq 0 \end{cases}$$

The coefficients are therefore defined by the equation: $a_n = \frac{D^n_{z=0}(f_N(z))}{n!}$, and the error is measured by the formula: $|E_n| \le \frac{\bar{M}k^{n+1}}{1-k}$, where $\bar{M}$is the maximum modulos of $f_N(z)$in the disc $|z|\le ka$, $k\in(0,1)$, suitably chosen so that the disc $|z|\le 1$is enclosed inside the disc $|z|\le ka$.

If Tdenotes the Taylor expansion, the overall error is expressed by the formula:

$$\left|\frac{1}{z}\right| \leq \sum_{i=2}^{N}(2^{n_{i-1}-n_i}+2^{-n_i})+2^{-n_1}+\frac{\bar{M}k^{n+1}}{1-k}.$$

Example

The function $f(\lambda)=\frac{1}{\sqrt{2\pi}}(1+\cos\lambda)$denotes the spectral density of a particular linearly regular MA(1) processes and then

$$g_\varepsilon(\lambda)=\frac{1}{\sqrt{2\pi}}(1+\cos\lambda)I_{[-\pi,-\pi+\varepsilon]\cup[\pi-\varepsilon,\pi]}(\lambda)$$

defines a class, for each $\varepsilon\in(0,\pi]$, of linearly singular processes.

Taking, for instance, $\varepsilon=\frac{3}{4}\pi$we shall have $\mathrm{dist}(K,[-a,0])=\sin(\pi-\frac{\pi}{4})=\frac{\sqrt{2}}{2}$with $a>1$if $a=2\frac{\sqrt{2}}{2}$.

Then, the rational function we are looking for is of type $f_2(\varepsilon)$(see paragraph 4 for notations) and the approximation error is

$$\left|\frac{1}{z}-f_2(z)\right| \leq 2^{-n_1}+2^{n_1-n_2}+2^{-n_2},$$

where the quantities are fixed depending on the error we want. By the paragraph 4, being $\alpha_1=\frac{1}{2}\sin(\frac{3}{4}\pi)=\frac{\sqrt{2}}{4}$and $\alpha_2=\frac{\sqrt{2}}{2}$, we have

$$f_2(z)=\left[\frac{(\frac{\sqrt{2}}{4}-z)^{n_1}-(\frac{\sqrt{2}}{4})^{n_1}}{z(\frac{\sqrt{2}}{4}-z)^{n_1}}\right]\cdot\left[1-\left(\frac{-\frac{\sqrt{2}}{4}}{z-\frac{\sqrt{2}}{2}}\right)^{n_2}\right]^{n_1}$$

and, taking $n_2=n_1=1$, the error is not grater than 2 and $f_2(z)=\frac{1}{z-\frac{\sqrt{2}}{2}}$.

The Taylor expansion in the origin of such function allows us to compute the predictor coefficients:

$$a_0=\frac{-2}{\sqrt{2}} \qquad a_j=\left(\frac{\sqrt{2}}{2}\right)^{-j}\frac{1}{j!}, \qquad j\neq 0$$

As a further example, choosing $n_1=n_2=2$, we obtain an error less than

$$2^{-2}+2^{0}+2^{-2}=\frac{3}{2}, \quad f_2(z)=\left[\frac{(z-\frac{\sqrt{2}}{4})(z-\frac{3}{4}\sqrt{2})}{(z-\frac{\sqrt{2}}{2})^2}\right]^2$$

and the predictor coefficients are

$$a_0 = \frac{-9}{2\sqrt{2}}$$

$$a_j = \frac{1}{j!}\left[\frac{3}{4}j(j+1)\sqrt{2}\left(\frac{\sqrt{2}}{2}\right)^{-(j+2)} - \frac{9}{16}(j+1)(j+2)\left(\frac{\sqrt{2}}{2}\right)^{-(j+3)} + \right.$$
$$\left. - \binom{j}{2}\left(\frac{\sqrt{2}}{2}\right)^{-(j+1)}\right], \qquad j \neq 0$$

Finally if n_1, n_2take any two non null values we have:

$$f_2(z) = \left[\frac{(\frac{\sqrt{2}}{4} - z)^{n_1} - (\frac{\sqrt{2}}{4})^{n_1}}{z(\frac{\sqrt{2}}{4} - z)^{n_1}}\right]\left[1 - \left(\frac{-\frac{\sqrt{2}}{4}}{z - \frac{\sqrt{2}}{2}}\right)^{n_2}\right]^{n_1}$$

and then the coefficients may be computed by the Taylor expansion. Furhtermore the error given by the quantity $2^{-n_1} + 2^{n_1 - n_2} + 2^{-n_2}$is decreasing to zero if $n_1 < n_2$and when $(n_1, n_2) \to (+\infty, +\infty)$.

For example: if $n_1 = 21$, $n_2 = 42$then $|f_2(z) - \frac{1}{z}| \leq 0{,}000002861$.

5.1 References

CELANT, G. (1983). Una dimostrazione sulla equivalenza tra processi stocastici debolmente stazionari a parametro discreto con traiettorie quasi periodiche e processi stocastici debolmente stazionari a parametro discreto con spettro a gradini, CNR, LADSEB 83.

CELANT, G. and LESSI, O. (1983). Asymptotic behaviour of the covariance function of some deterministic components in weakly stationary stochastic processes. European young statisticians meeting, Leuven, 22-26 August 1983.

CELANT, G. (1985). Proprietà spettrali e comportamento asintotico dei processi stocastici debolmente stazionari Cantoriani, SIS, Convegno nazionale di metodologia statistica, Bressanone 18-20 Settembre 1985.

CELANT, G. (1988). Alcuni problemi statistici relativi ai processi singolari spaziali. Statistica, anno XLVIII, n. 1-2.

IBRAHIMOV, I. and ROZANOV, Y. (1974). Processus aléatoires gaussiens, MIR.

LAURENTIER, M. and CHABOT, B. (1977). Méthodes de la théorie des fonctions *d'* une variable complexe, 2e édition, MIR.

PRIESTLEY, M.B. (1981). Spectral Analysis and Time Series, Vols. 1 and 2, Academic Press, New York.

SCHAEFER, H.M. (1980). Introduzione alla teoria spettrale, Boringhieri.

GPSR Compliance

The European Union's (EU) General Product Safety Regulation (GPSR) is a set of rules that requires consumer products to be safe and our obligations to ensure this.

If you have any concerns about our products, you can contact us on ProductSafety@springernature.com

In case Publisher is established outside the EU, the EU authorized representative is:

Springer Nature Customer Service Center GmbH
Europaplatz 3
69115 Heidelberg, Germany

Batch number: 10725973

Printed by Printforce, the Netherlands